HIGH TECHNOLOGY, SPACE, AND SOCIETY

Volume 28, URBAN AFFAIRS ANNUAL REVIEWS

HIGH TECHNOLOGY, SPACE, AND SOCIETY

Edited by
MANUEL CASTELLS

Volume 28, URBAN AFFAIRS ANNUAL REVIEWS

SAGE PUBLICATIONS
Beverly Hills   London   New Delhi

Copyright © 1985 by Sage Publications, Inc.

For information address:

SAGE Publications, Inc.
275 South Beverly Drive
Beverly Hills, California 90212

SAGE Publications India Pvt. Ltd.
M-32 Market
Greater Kailash I
New Delhi 110 048 India

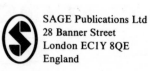

SAGE Publications Ltd
28 Banner Street
London EC1Y 8QE
England

Printed in the United States of America

Library of Congress Cataloging in Publication Data

High technology, space, and society.

    (Urban affairs annual reviews ; v. 28)
    1. Technology—Social aspects—Addresses, essays,
lectures. 2. Space in economics—Addresses, essays,
lectures. I. Castells, Manuel. II. Series.
HT108.U7   vol. 28   307.76 s [303.4′83]   84-24876
[T14.5]
ISBN 0-8039-2415-1
ISBN 0-8039-2414-3 (pbk.)

SECOND PRINTING, 1987

Contents

Preface

☐ TWO MAJOR PHENOMENA are taking place in our societies: a technological revolution of extraordinary proportions and far-reaching implications and a major process of urban-regional restructuring that is reshaping the spatial forms and dynamics at the world level. The simultaneity of these two processes has led numerous observers and policy makers to assume a causal relationship between them, with technology as the leading force of human progress, to whose requirements and logic cities, regions, and nations must adapt.

Such a view has obvious social and political implications, particularly in a period of structural economic change. In fact, historical reality is much more complex, and the scarce reliable evidence on this issue seems to point to the existence of a web of interactions between technology and space, mediated by economic, cultural and political processes. This volume explores such relationships on the basis of a series of original research essays whose only common ground is the recognition of the importance of technological change for the evolution of spatial and social forms, *together with an emphasis on the need to integrate technology in a broader framework of social relationships* to understand the diversity of its effects on people's lives, on institutions, and, ultimately, on spatial forms and processes. Beyond this common approach and interest, the analyses and hypotheses presented in this volume are deliberately diverse, because such is the state of research on this topic at the current exploratory stage of its development. It is particularly important to note that the overview chapters by Castells and by Hall do *not* represent a synthesis of the contributions to the volume, as they have been produced in a parallel process, without previous knowledge of the other chapters. It is the purpose of this text to present a sample of research currently under way on this new frontier of urban studies, without attempting the construction of a coherent view of the phenomenon. As a matter of fact, the reader will be interested, and maybe amused, by the open empirical and theoretical contradictions between some of the chapters of the volume. It is my opinion that such debates are the only fruitful way to proceed in an area of research in which the importance of the issues tends to favor prophetic statements more often than scholarly research. And yet we need, first of all, to establish a relatively solid ground, both empirical and theoretical, before proposing further generalizations. In particular, we have not attempted a more precise definition of the nontheoretical notion of "high technology," keeping it in its commonsense, journalistic meaning. Although several chapters in the volume try to provide a more analytical or a more operational delimitation of high technology, we have preferred to leave open its meaning to the actual content provided by the research presented here. Thus, we do not start with a definition of high technology. We proceed, instead, toward an approximation to the understanding of the processes to which "high tech" debates generally refer, through the research work gathered in this volume.

The organization of the book, while relatively coherent, does not cover (and could not cover) all important issues. The final outline is highly dependent upon the availability of original research material in this very new field of study. Thus, after an attempt to summarize some of the existing information on the United States and Great Britain within two particular analytical perspectives, the volume presents some of the most complete empirical findings to date on the locational patterns and regional effects of high technology manufacturing, pinpointing some of the areas (Silicon Valley, Route 128), and some major issues (the military connection) of the new industrial space. We proceed, then, to present research on the technological component of the transformation of services, as it relates to the metropolitan economies, with particular emphasis in one chapter on the effects for working women.

Next, the volume examines the changes in the field of communications, focusing on the emerging "on line economy," and on the social, spatial, and political impacts of the new media. Three chapters present theoretical approaches to the study of technology, space, and society, introducing both a word of caution and some critical perspectives to the research-subject itself. Finally, alternative views are expressed concerning social and spatial uses of new technologies that could embody a different, more humane set of values.

The very openness of the volume invites to its development in the practice of a new generation of urban researchers to whose ability to supersede the current shortcomings of our work we would like to have contributed with this collective effort.

—Manuel Castells
Berkeley, California

Part I

Overviews

1

High Technology, Economic Restructuring, and the Urban-Regional Process in the United States

MANUEL CASTELLS

□ WE ARE IN THE MIDDLE of a major technological revolution that is transforming our ways of producing, consuming, organizing, living, and dying. Cities and regions are also changing under the impact of new technologies, To be sure, technology does not develop in a social vacuum, and its effects are necessarily dependent upon a broader framework of economic, cultural, and political processes, some of which I will try to identify in this chapter. Yet it is important to analyze the specificity of technological change in order to understand the novelty of the current process of transformation of the spatial structure.

Two features are characteristic of the stream of technological innovation under way. First, the object of technological discoveries, as well as of their applications, is *information*. What microelectronics does is to process, and eventually generate, information. What telecommunications do is to transmit information, with a growing complexity of interactive loops and feedbacks, at increasingly greater speed and at a lower cost. What the new media do is to disseminate information in a way potentially more and more decentralized and individualized. What automation does is to introduce preinformed devices in other activities. And what genetic engineering does is to decode the information system of the living matter and try to program it.

The second feature concerns the fact that the outcome is *process-oriented,* rather than *product-oriented.* High technology is not a particular technique, but a form of production and organization that can affect all spheres of activity by transforming their operation in order to achieve greater productivity or better performance, through increased knowledge of the process itself.

These technologies interact with the spatial structure in three fundamental ways: (1) The new informational logic of production and management creates *a new space of production,* the development of which fundamentally reshapes the regional structure and the dynamics of each city, according to the functional importance of their social, economic, and institutional characteristics for the new production system. (2) *The direct impact of new technologies* (particularly of communication technologies) on the ways of working and living *tends to modify the urban form.* (3) Yet the effects of high technology are mediated by broader social and economic processes that frame its uses. *The most important global process conditioning the relationship between new technologies and spatial dynamics is the economic restructuring that U.S. capitalism is currently undergoing, superseding the structural crisis of the 1970s.*

Thus, I will subsequently analyze these three forms of interaction between high technology and space, on the basis of available empirical evidence on U.S. cities and regions.[1]

A NEW SPACE OF PRODUCTION

The most direct impact of high technology on the spatial structure concerns the emergence of a new space of production as a result of two fundamental processes: on one hand, high technology activities become the engine of new economic growth and play a major role in the rise and decline of regions and metropolitan areas, according to their suitability to the requirements of high tech production; on the other hand, the introduction of new technologies in all kinds of economic activities allows the transformation of their locational behavior, overcoming the need for spatial contiguity.

On the basis of recent empirical research on the spatial behavior of industrial sectors (Glasmeier et al., 1983; Storper, 1982; Pinkerton, 1984; Walker and Storper, 1984; Mutlu, 1979; Saxenian, 1980, 1984), a model of location for high tech manufacturing can be proposed. Such a model requires the combination of five characteristics to make a given space attractive for high tech activities. The five characteristics are determined by the specificity of the processes of production and management in high technology manufacturing. The proposed characteristics are as follows:

(1) Because high tech industries are science-based and knowledge-intensive, they need a close connection to major universities and research units, as well as to a large pool of technical and scientific labor (Borrus and Millstein, 1982; Hall et al., 1983; Sylvester and Klotz, 1983).

(2) Given the dependence on government markets, particularly on the military and space programs and especially until the late 1960s, high tech activities tend to cluster historically in regions where the military has established its testing sites (Mutlu, 1979; Storper, 1982; Carlson and Lyman, 1984). This explains why defense spending is the variable that in all statistical studies most often correlates with high tech location (Glasmeier et al., 1983; Markusen, 1984b; Pinkerton, 1984).

(3) High tech companies are generally characterized by a strong antiunion feeling in their managements. Not so much because of traditional economic reasons such as wages or benefits, but because of the fears of bureaucratization and slowness in an industry that requires constant flexibility and innovation (Rogers and Larsen, 1984; Saxenian, 1980). Thus, areas with a strong union tradition will tend to discourage high tech location, all other elements being equal.

(4) The risk (and promise) of investment in this new field requires the existence of venture capital in the region, that is both a function of a high level of wealth and of an entrepreneurial culture oriented toward nontraditional financial markets (Mutlu, 1979; Borrus and Millstein, 1982; Carlson and Lyman, 1984).

(5) The process of production in high tech in general, and in microelectronics in particular, is highly discrete and can be easily separated in time and space between its research and design, fabrication, assembly and testing functions. Given the very different requirements of each function, especially for labor, it follows that there is a heirarchical division of labor across space, and the need for all activities to be located in a good position in a communications network (Mutlu, 1979; Saxenian, 1980; Saxenian, 1984; Storper, 1982).

This model explains high tech location better than the vague opposition between the snowbelt and the sunbelt (invalidated by New England), or the very subjective notion of the "quality of life" (certainly challenged by hardcore New Yorkers if it only evokes sunny California or colorful Texas). For instance, Santa Clara county ("Silicon Valley") possesses all of the five characteristics cited; the second high tech nest (the first until the mid-1960s), Route 128 (Boston), has almost everything, but it was somewhat limited in development by its strong union influence and the lack of military settlements. Texas and Arizona occupy an intermediate position, largely due to the absence of major research centers (Austin is now in the process of reversing the trend) and, to some extent, to the fact that their financial markets were less well-structured than those in San Francisco or Boston. New York-New Jersey offers the counterexample of how the concentration of giants such as IBM or RCA could not generate a milieu of innovation outside the companies themselves in spite of such assets as Bell Laboratories, Princeton, Cold Spring Laboratory, the largest financial center in the world, and the proximity to the largest information market. The absence of the military factor and the prounion environment in New York were powerful obstacles to the development of a high tech milieu, while the large firms such as IBM were actually quite happy in building their empire inward, in isolation of potential competition and in prevention of threatening spin-offs (Mutlu, 1979).

The current sprawl of high tech activities modifies the geography of their location, but not the logic of their spatial relationships, still based on the hierarchy of intracompany units and of areas between themselves.[2] New high tech centers have developed (in Southern California, North Carolina, Texas, Colorado, Utah, Washington, Arizona, Pennsylvania, Florida, and Maryland) along with three complementary tendencies for the location of assembly lines: off-shore production (Pacific Rim, Mexico); isolated rural communities in the Western United States (Oregon, Arizona, Colorado, Texas); and, recently, fully automated plants closer to the high tech nests (Glasmeier, 1985; Saxenian, 1980; Glasmeier et al., 1984; Henderson and Scott, 1984). Given the growth potential of these activities, and even more, the myth surrounding their generative character, cities and regions have entered the race to attract them, thus fundamentally changing the economic geography of the entire nation.

The new space of production results also from the new functional linkages allowed by communication technologies (Nicol, 1983a; Schement et al., 1983). Unlike what is often argued, the increasing "footlooseness" of many activities does not result in the suppression of the spatial dimension of the economy or of society (Nicol, 1983b). What happens is that the internal logic of organizations *tends* to supersede the external logic of factors associated with specific places (Nicol, 1983a; Stanback, 1979).

We still have to deal with space, but we increasingly observe *a space of flows substituting a space of places* (Castells, 1983a; Martin, 1981). Thus, a hierarchy of functions and power positions structures the territory across the nation and across the world, separating functions and units of production, distribution, and management to locate each one in the most favorable area, yet articulating all activities through a communication network (Friedmann and Wolff, 1982; Webber, 1980). As the organizational logic changes often, and as the social and economic system responds to a multiplicity of large-scale organizations, we are living increasingly in a space of variable geometry where the meaning of each locale escapes to its history, culture, or institutions, to be constantly redefined by an abstract network of information strategies and decisions. The new space of production and management is *not* the result of new communication technologies, but it is through the possibilities offered by them that it comes into being historically.

TOWARD THE INFORMATIONAL CITY?

Technological innovation has always played a major role in the shaping of urban forms. It is likely to be even more so during the current process of discovery and implementation of new communication technologies. Nevertheless, the impact of these technologies on the city will be mediated and fundamentally modified by economic, social, and cultural processes. This is why, when discussing this particular issue, we must consider the technological impacts as *tendencies* rather than as direct effects, to create some distance between our analytical perspective and the superficial prophecies of many "futurologists."[3]

In the sphere of work widespread *office automation* (Baran, 1982; Strassman, 1980), improved *data transmission* (Dertouzos and Moses, 1981), cable and *telecommunications* in general (Baldwin and McVoy, 1983) offer the possibility for a generalized decentralization of the work place, and even for "cottage work" (see EDP Analyser, 1982). Some estimates consider that within a decade 18% of the U.S. workforce will be *telecommuting* (Business Week, 1984a). In fact, most likely this trend will be limited to those professionals who are already doing some work at home, besides going to their offices on an irregular schedule. Yet, because of the growing quantitative importance of this occupational group in the overall economy, the phenomenon could be sizeable in the near future. The possibility for unskilled clerical work to be done at home or in decentralized office units already exists on technical grounds (particularly if telephone lines are massively converted to optical fibers and digital transmission), but its development is still very limited and will depend, fundamentally, on the evolution of labor-management relations in the office of the future: Namely, if clerical unionization grows, companies will have a strong incentive to stimulate isolation and part-time work with a flexible schedule and remote control of routine operations. If such a trend develops, the "office of the future" could resemble the "domestic sweatshops" of the first period of industrialization. Yet the process of technological transformation of the office is complex and contradictory. Research by Barbara Baran and Suzanne Teegarden on the insurance industry has shown that two opposite trends develop at the same time (Baran and Teegarden, 1983; Baran, 1984): regional and metropolitan decentralization of standard data-processing operations, on the one hand; reintegration of different functions in the activity of a multifunctional, highly qualified work team in a central location, on the other. Thus, the transformation of the spatial pattern of service activities will depend on the logic of these activities, and will not just be a function of the new technologies. What technology allows is that the organization locates itself according to its strategy, and not because of its spatial constraints. In general, given the high cost of location in most CBDs, the tendency is toward increasing territorial sprawl of services, both to regions distant from the location of major headquarters and to the suburbs of the metropolitan areas (Daniels, 1979; Stanback, 1979). The electronic highways of the informational city do not substitute for the current transportation network: They actually extend the spatial area of activity, and increasingly diffuse the urban subsystems in a global territorial structure.

Nevertheless, the most important consequence of new communication technologies for urban life and forms is what has been labeled the "*home information revolution*" (Williams, 1982). Many households in the United States are becoming real communication hubs. By the end of 1983, about 7% of U.S. households were equipped with home computers, 12% with video-cassette recorders (VCRs), 21% had programmable video games and 37% subscribed to some kind of Cable TV (Business Week, 1984b; New York Times, 1984). To the 150 million television sets in use in the United States, millions of VCRs have now been added. In 1983 10 million households had at least one VCR, in 1984 that figure was expected to rise to 15 million, and the projection was about 35 million in 1988, which is more than one third of U.S. households (Newsweek, 1984; Stark, 1984). Along with the video explosion, we observe the development of new forms of TV (high-definition TV, stereo TV, small 13-inch color TV, giant screens, and, most importantly, digital TV with perfectly clear images, able to hold the picture, divide the screen, "zoom in" parts of it) and soon, specialized radio, portable audio-visual equipment, and affordable roof-top discs able to receive hundreds of channels through satellite transmission (Moorfoot, 1982; Rice, 1984; Mahony et al., 1980).

Yet if we know what the newest developments are, it is most uncertain what is (or will be) the actual use of this equipment. Word processing seems to be the overwhelming use for personal computers, which is limited to a minority of professionals. Games and tax accounting are the other most frequent uses, but they certainly do not justify full-time possession of a computer. It is in the field of *on-line information services* that the new electronic home could take off more rapidly, particularly through the telephone (Dordick et al., 1979; Martin, 1981; Williams, 1982). While it is doubtful that people will like electronic mail, electronic banking, or teleshopping, it is certain that there is a drive from large organizations to save time and labor costs by stimulating households to hook up into their systems (Calhoun, 1981). This will be followed by a tendency to increase the number of services and information transmitted at distance, with interactive flows through cable or telephone. Thus, a slow but steady trend toward the "telecommunicated city" or the "wired city" does appear (Webber, 1980). The likely impact will be a decrease in functional traveling around urban areas and a concentration of activities around three major poles: work places, homes, and pure leisure places. This trend increases the functional zoning of time and space in such a way that public space is being reduced to the space of leisure (for those who have the time and money) and to the space of wandering (for those who do not fit in the functional assignment of work and residence). Technology assists to a new step toward the disintegration of urban cultures that were always characterized by the spatial mixture of uses. Land use zoning started to segregate people and activities. Now "electronic zoning" dramatically enhances this tendency, transforming places into unifunctional units.

Nevertheless, the real "revolution" occurring at home is in entertainment (Stark, 1984). Homes increasingly are becoming equipped with a self-sufficient world of images, sounds, news, and information exchanges. Video recording reinforces this self-sufficiency with the possibility of greater freedom for the time and content of the viewing. Homes at the same time become instant receivers of planetary information and personal refuges of selective consumption of images and sounds. Furthermore, the trend is not only toward home-centered entertainment, but toward the individualization of the communication experiences, with highly specialized radio stations and portable audio and visual equipment as the clearest expression of this new trend (Williams, 1982). Thus, not only will people be able to stay at home, seeing nobody and yet receiving news from the entire world, or filling their eyes and ears with a whole realm of experiences, but they will also be able (as they already *are*) to leave home without leaving their inner experience. Thus, as a *tendency,* we can say that the new technologies lead toward the delocalization of experience in the sphere of private life, as they do for work-oriented organizations. Homes *could* become disassociated from neighborhoods and cities and still not be lonely, isolated places. They would be populated by voices, by images, by sounds, by ideas, by games, by colors, by news. And yet you (we) could switch it all off in one gesture.

So, we know that the world (at home) is only there if we wish so. From a strictly technological point of view, there is no more mediation between the individual and a global culture, satellite-transmitted, then specialized and targeted to specific people and to specific moods. Thus, in between, there would be no more society and no more city. At least not the ones we have known. Nevertheless, social relations are *not* the direct result of technological change, so the emerging urban form will be shaped also by other processes; for instance, by the persistence of ethnic cultures, local networks, and community organizations. But when we know the increasing proportions of people living alone (about 25% of households in the United States in 1980), the decreasing size of households, and the crisis of the nuclear family (in 1970 70.5% of total households were maintained by a married couple, whereas in 1982 the proportion had decreased to 59.4%, Norton, 1983), it appears that the current technological trends could reinforce the social tendency toward individualization.

On the other hand, along with territorial sprawl, metropolitan decentralization, and individualization of the residential landscape, the new technologies also enhance, simultaneously, the importance of a few places as locations of those activities that cannot be easily transformed into flows and that still require spatial contiguity, thus reinforcing considerably the intraurban hierarchy. In the informational city, spatial singularity and urban centrality become even more important than in the industrial-commercial city, precisely because of their unique locational requirements. Thus, high-level managerial functions, specialized leisure areas, key informational institutions, very specific production centers (such as high tech nests), and special service-delivery activities (from hospitals to high-class boutiques) will still earmark the metropolitan space with their requirements for spatial contiguity and face-to-face interaction. Thus, spatial indifferentiation and extreme specificity of a few nodal places are both characteristics of the informational city foreshadowed by the new communication technologies. Besides, these new spatial flows do not develop in a historical vacuum. They take shape on the structures of the preceding urban civilization and cut across the old and new situations of social exclusion. Flows connect networks that are functionally useful and socially valued (Baldwin et al., 1980). Nodal places nest the most important activities and welcome the new residential elite. Secluded individualistic homes across an endless suburban sprawl turn inward to preserve their own logic and values, closing their doors to the immediate surrounding environment and opening their antennas to the sounds and images of the entire galaxy. The issue arises that in such a structure organized around flows, people, activities, and cultures that are not valued (or priced) could easily be switched off the network (Williams, 1982). And in a city where the only meaningful places are the ones associated with the highest functions, the space with meaning for only a few tends to be the space of exclusion for the most. Thus, between the discontinuous spatial elements of the informational city, there will remain switched-off, wireless communities, still real people in real places, yet transformed into urban shadows doomed to haunt the ultimate urban dream of the new technocracy.

ECONOMIC RESTRUCTURING AND THE URBAN-REGIONAL PROCESS

The essential impact of the current technological revolution is being felt in the economic sphere. And because economic factors fundamentally affect the spatial structure, it is one of my major hypotheses that it is mainly through the new economy that high technology is deeply modifying our cities and regions. For the sake of clarity, the method of analysis attempted here will be to identify the role played by the new technologies in each of the major processes at work in the current restructuring of U.S. capitalism. I will then elaborate on the consequences of each of these processes, with their technological component, on the urban-regional structure. This perspective assumes that the world economic system, still centered in the United States, is in a process of deep structural transformation, in intimate connection with the technological revolution (although *not* determined by it), and that only by understanding the effects of technology within this broader framework (we call it the economic restructuring of the world capitalist system) will we be able to assess the impact of high technology in the new urban-regional process.

As I have argued elsewhere (Castells, 1980), the capitalist system went through a major crisis in the 1973-1982 decade (Carnoy et al., 1983; Bowles et al., 1983; O'Connor, 1984) and is emerging from it by means of setting up a new model of economic accumulation, social organization, and political legitimation that, while still being capitalist, will be as different from the Keynesian model of the 1945-1973 period as the latter was different from pre-Great Depression capitalism (Carnoy and Castells, 1984). Such a model relies on three main processes. The first is a fundamental *transformation of the capital-labor relationship in the work process,* with capital gaining the initiative again over the wages and regulations conquered by the labor movement after decades of class struggle (Gordon et al., 1982; Portes and Walton, 1981; Sabel, 1982; Carnoy and Shearer, 1980). The second is *a new role of the state and of the public sector,* not so much reducing government intervention in the economy, but shifting its emphasis from collective consumption to capital accumulation, and from legitimation to domination, something that can be schematically called the transition from the Welfare State to the Warfare State (O'Connor, 1973; Gough, 1979; Wilensky, 1974; OECD, 1981; Crouch, 1979; Leontieff and Duchin, 1983; Dumas, 1982). And the third is *a new international and interregional division of labor,* with capital, labor, production, markets, and management shifting locations in a continuing variable geometry to take advantage of the best possible conditions for the strategy of the large corporations, regardless of the social and political consequences for specific territorial units (Frobel et al., 1980; Bluestone and Harrison, 1982; Sawers and Tabb, 1984; Palloix, 1977).

I propose as a hypothesis that the new technologies are playing a major role in these three fundamental processes and that such a technological-economic impact is crucial to understanding the emerging spatial structure. What I attempt in this text is to explore the urban-regional profile of a new socioeconomic model at the early stage of its implementation.

HIGH TECHNOLOGY, LABOR, AND CITIES

The first process of economic restructuring concerns the deterioration of the wages and working conditions obtained by labor over capital after hard social struggles for many decades. Given the substantial share of labor costs in total production costs, the new strategy of cheapening labor and weakening union power is a paramount goal for corporations. The new technologies play a major role in the reorganization of the labor process to implement this strategy along four major lines:[4]

— Automation in both factories and offices allows labor-saving procedures that dramatically increase productivity while eliminating jobs on a massive scale, particularly in manufacturing and in heavily unionized sectors.

— By creating a permanent threat of substituting machines for workers, management puts a tremendous pressure on labor, which will be forced to accept, in many cases, whatever conditions are considered profitable by capital.

— Automation of information-processing functions allows capital to supersede one of the major barriers for the growth of economic productivity. The expansion of the service sector in employment at a much lower rate than its contribution to the GNP led to a structural contradiction that could be one of the sources of the slowing of productivity growth in the United States: On one hand, producer services (and to a lesser extent social services) have become crucial for stimulating productivity in the overall economy. On the other hand, these service activities continued to be labor-intensive and prone to low productivity growth (Stanback, 1979; Mark and Waldorf, 1983; Denison, 1979). The possibility of office automation allows corporations to grow in size without losing flexibility and efficiency, and make it possible to boost productivity in the management process, potentially overcoming the obstacles to labor productivity in the service sector (Chandler, 1977; Cohen, 1979). Yet, as services were the refuge for the growing labor surplus from other sectors as well as a response to the political pressures on the public sector to provide jobs and services, automation of services will be uneven and will provoke bitter social conflicts (Singleman, 1977; Castells, 1979).

— Scattered observations on the impact of automation on the occupational structure seem to indicate that it will result in a bifurcated labor market, with the upgrading of a minority of workers and rapid growth of professional sectors, while a majority of workers are deskilled and reduced to low-paying jobs, either in labor-intensive services or in down-graded manufacturing (Markusen, 1983; Hirschorn, 1984; Storper, 1982; Glasmeier, 1985). This bifurcated occupational structure reflects itself in the income distribution, with a tendency to what Lester Thurow calls "the disappearance of the middle class" (Thurow, 1984).

Nevertheless, if the massive elimination of jobs in traditional activities due to the introduction of new technologies is recognized by most experts,[5] the question is the potential for growth and jobs generated in the new high tech sectors. In fact, they do generate jobs, but not enough to compensate for the ones eliminated by technological change. Ann Markusen (1983) estimates that during the 1980s high tech activities will generate 3 million new jobs, but their uses will eliminate 25 million jobs. In this sense, it is important to distinguish between rates of growth and actual aggregate growth in employment. For instance, between 1982 and 1990 the Bureau of Labor Statistics projects a 27% growth in computers and peripherals, but total employment in the sector will be about 600,000 jobs in 1990. While the moderate 15% growth in "restaurants and other retailing" will take the number of jobs in this sector to 20 million in 1990. Thus, high tech will eliminate many more jobs than it will create by itself. Yet it still could be argued that the overall dynamics of the economy, helped by high technology, will generate new jobs outside high technology activities in new sectors different from those being automated. And, in fact, such seems to be the current trend in the United States (Wayne, 1984). As a point of comparison, during the crisis of the 1970s, while Western Europe lost almost 3 million jobs, the American economy created *20 million new jobs*, the factor accounting for the difference being, most likely, the much greater power of unions and socialist parties in Europe, which made business reluctant to invest in job creation because of high labor costs. Given the slowing of the rate of growth of the U.S. workforce in the 1980s (1.5% annual rate, against 2.7% in the 1970s), the political acceptance of 7%-8% unemployment as "normal," and the expansion of the "underground economy," the impact of technology on labor is likely to be socially assimilated in terms of numbers of available jobs. Nonetheless, the fundamental feature is the mismatching between the characteristics of jobs that disappear and those of the newly created jobs. On one hand, an upgrading of professional and technical jobs in advanced services and high technology manufacturing requires a fundamental retraining of labor, something the educational system is hardly able to assure, particularly in the secondary public school. On the other hand, the new unskilled jobs are to be found in low-paying services and down-graded manufacturing, where cheap labor is cheaper than automation, where companies cannot afford capital investment, or where the nature of the activity makes it difficult to standardize work (Bluestone and Harrison, 1982; Serrin, 1983). So routine service jobs, new "sweatshop manufacturing," and nonunion labor are substituting for the traditional unionized automobile or steel worker, for the government employee, or for the stable clerk. The mismatching between old and new jobs has been documented by Bluestone and Harrison (1984). According to their study, in 1982, when one compares weekly wages from declining sectors to expanding sectors, one finds a reduction from $310 to $210 (in the expanding sectors). Similarly, "it takes almost 2 department store jobs or 3 restaurant jobs to make up for the earnings loss of just one average manufacturing position" (Bluestone and Harrison, 1984:25). Other studies find similar trends in the new labor market (Walker, 1983; Storper, 1982; Hirschorn, 1984; Teitz, 1984; Tomaskovic-Devey and Miller, 1982)

The restructuring of the labor process, and thus of labor markets, by new social relationships and high technology, is having a major effect on the urban-regional structure at four levels. First, the traditional manufacturing regions and cities where old line factories were located are being abandoned or automated, with considerable decline and outmigration of jobs (Hicks and Glickman, 1983; Mollenkopf, 1983; Noyelle and Stanback, 1984). Second, new regions emerge as manufacturing centers: those where high tech nests are located; those where labor costs are low and the socio-political environment is "good for business"; and some isolated rural areas where new high tech industries decentralize assembly-line operations.

Also, expansion of advanced corporate services, supported by office automation, is concentrating companies, headquarters, and their constellation of auxiliary services (both producer and consumer services) in the downtowns of the top metropolitan areas, many of which are literally booming (Cohen, 1978; Mollenkopf, 1983; Strassman, 1980; Stanback et al., 1981), thus stimulating a surge in construction of new high-rise buildings as well as rehabilitation and conversion of old structures (Mollenkopf, 1984). Paradoxically, the decline in public urban services is hampering the perspective of further downtown development.

And, finally, an even more fundamental spatial restructuring is under way, directly connected with the new capital-labor relationship facilitated by high technology: *the new economic and social dualism within the largest metropolitan areas,* as has been shown by empirical studies of the recent evolution of New York (Mollenkopf, 1984; Sassen-Koob, 1984b), and Los Angeles (Soja et al., 1983; Sassen-Koob, 1984a). Several processes are taking place simultaneously within the same metropolitan area: rapid growth of advanced corporate services and high tech manufacturing; decline of traditional activities (both in services and manufacturing); and development of the new downgraded, yet booming, economic sectors ("sweatshops" with undocumented workers, eating and drinking places, luxury consumption, and so on).

This process of *polarized growth* creates distinct social spheres, yet it has to link these spheres within the same functional unit. This trend is different from the old phenomenon of social inequality and spatial segregation in the big city. There is something else than the distinction between rich and poor or white and nonwhite. It is the formation of different systems of production and social organization, which are equally dynamic and equally new, yet profoundly different in the wealth, power, and prestige that they accumulate. The new immigrant neighborhoods of New York or Los Angeles are lively areas, not rundown ghettoes, in spite of the misery and oppression experienced by many of their dwellers. Their role in the new metropolitan economy is crucial for both production and consumption and their cultural autonomy transforms the life and politics of the city. Yet these different worlds (the high tech world, the advanced services world, the auxiliary services world, the various immigrant worlds, the traditional black ghettoes, the protected middle-class suburbs, etc.) develop along separate lines in terms of their own dynamics, while still contributing altogether to the complex picture of the new supercity. The new labor market is at the basis of this newly polarized sociospatial structure. We witness the rise of dualized supercities that segregate internally their activities, social groups, and cultures while reconnecting them in terms of their structural interdependency. These metropolises are magnets on a world level, attracting people, capital, minds, information, materials, and energy while keeping separate the channels of operation for all these elements in the actual fabric of the metropolis. We are not in a situation of urban-regional crisis (as it could be in Detroit or Buffalo), but in a process of interactive growth between elements that ignore one another while being, in fact, part of the same system. We are witnessing the rise of urban schizophrenia. Or, in other words, the contradictory coexistence of different social, cultural, and economic logics within the same spatial structure.

HIGH TECHNOLOGY
AND THE SPATIAL FORM
OF THE WARFARE STATE

The most far-reaching element of the current economic restructuring is the transformation of the role the state has held in the U.S.

economy for the 50 years since the New Deal (O'Connor, 1973; Janowitz, 1980; Carnoy, 1984; Wolfe, 1976). Not that it is withdrawing from economic interventionism, quite the opposite. But the form and content of government intervention are deeply changing, shifting their focus from accumulation and redistribution, to *selective* accumulation and military reinforcement. It would be a mistake to consider this trend a product of the Reagan Administration, although its policy represents a qualitative step in this direction (Wilmoth, 1983; Leckachman, 1981; Palmer and Sawhill, 1982). In fact, the crisis of the Welfare State results from the method chosen to fight inflation: to cut back social expenditures, thus limiting the budget deficit and the need for public indebtment and excessive money supply (Gough, 1979; Adams and Freeman, 1982; OECD, 1981). Yet, as everybody knows, the budget deficit has skyrocketed in recent years because of the military build-up and the interest payments required by massive borrowing. So we witness *simultaneously* the partial dismantlement of the Welfare State, particularly in the sphere of urban programs and community services (Hirschorn et al., 1983), and the rise of the Warfare State. That is, we see the rise of a gigantic investment in military technology and warfare systems, to prepare for all kinds of wars on a planetary (and even a "galactic"!) scale (Dellums, 1983; Dumas, 1982; Gansler, 1980).

While the reasons for such trends are primarily political (as are all processes related to the state) their impact on the economy is decisive (Carnoy and Castells, 1984). Not only has the budget deficit gone out of control, but the economy as a whole is becoming increasingly militarized, reversing the trend of the 1960s and 1970s, when dependency from military expenditures was decreasing. The current economic dynamism is being restructured around a core of highly profitable industries related to military production, both for domestic consumption and for weapons export.

High technology is directly connected to the militarization of the economy in the United States. Again, not because technologies by themselves are military-prone (as the development of Japanese high tech oriented toward consumer electronics shows). In fact, even in the United States, after an initial period (particularly during the 1950s) when military markets overwhelmingly dominated the industry, particularly in electronics, during the 1970s consumer products and industrial applications became the most important part of high tech activities, as evidenced by the declining share of government in the semiconductors market. Nevertheless, during the 1980s the trend has been reversed and high tech manufacturing, particularly microelectronics, has been greatly stimulated by the military and space programs. A study by Markusen that was based on several data sources, including her own, concludes that "heavily defense-dependent manufacturing sectors are dominated by those in the high tech category, and a substantial proportion of high tech industries are military suppliers" (1984:27). She estimated that in 1977 military-related high tech production accounted for 47% of all high tech manufacturing employment. Given the Reagan military built-up, it is likely that this proportion increased considerably in recent years. Other empirical studies also show a close link between high tech production and military procurement, in terms of both employment (Henry, 1983; Dempsey and Schurde, 1971; Rutziek, 1970; Markusen, forthcoming) and the location of high tech industries (Clayton, 1962; Glasmeier, 1985; Pinkerton, 1984). In any case, what experts agree on is the decisive role played by the military in the origins of high technology developments (Carlson and Lyman, 1984; Mutlu, 1979). The emphasis on performance, regardless of the cost, the generous funding for long periods of time, the willingness to support innovation outside the large companies, and the requirement of miniaturization of electronic devices (to be used in airplanes and mobile supports), were all factors that facilitated the electronics revolution and shaped its products, even if many of them were developed afterward for all kinds of civilian applications.

These origins of military-related high tech production, along with the current trends of renewed militarization, have had decisive effects in the urban-regional process of the United States. While the dismantlement of the Welfare State has led to the urban fiscal crisis of most inner cities, the rise of the new Warfare State has spurred growth in new regions: in the suburbs of some metropolitan areas and among professional-technical sectors of a few cities. Norman Glickman (1983) has attempted to measure this impact by calculating the rates of growth and decline of federal outlays for selected programs between 1981 and 1984, for different cities and regions, typologized by a number of key variables. All welfare-related programs decline substantially, while defense spending increases sharply, particularly in the chapter "procurement." Interurban and interregional variations of social and defense expenditures follow almost systematically opposite patterns. Defense spending increase is more pronounced than average in high-income, nondeclining, low-unemployment, and "low-hardship" areas; in medium and small, rather than large, cities; and is overwhelmingly concentrated in suburbs, although this last result (interestingly enough) is due exclusively to the high suburban concentration of defense procurement. In terms of regions, while concentration ratios vary across the country depending on programs, defense expenditures are heavily concentrated in the West, South, Central, South Atlantic, Mountain, and especially in the Pacific regions. The study by Glasmeier et al. (1983) finds similar trends with high technology defense-related activities: they are the most spatially concentrated of all high tech sectors, and show a preference for the Southwest and the West, for medium-size cities, and for suburban locations. The reasons for this regional-urban location pattern seem to be functional, historical, and cultural (Markusen, 1984b). The military needs large extensions of isolated, undeveloped land, both for construction of huge facilities and for testing sites; it needs good weather all year round for aircraft and missile testing; it was located in the Pacific rim during the build-up against Japan in World War II; it prefers self-contained residential communities close to the production facility, so that the political environment will not jeopardize the continuity of production and the shared promilitary values, given the considerable cross-over between the army and the management positions in the military industry. Also, there is an antibig-city, antiurban culture that characterizes many of the military personnel, eager to live within their own world, far away from the social and political complexities of large inner cities. Thus, semirural regions and isolated suburbs in the metropolitan areas in the western United States seem to fit the cultural pattern of this "frontier spirit" of the military and related industries.

The Welfare State was born in the furnaces of life of the large inner cities. The Warfare State expands over an open space of distance and silence.

HIGH TECHNOLOGY
AND THE NEW SPATIAL
DIVISION OF LABOR

Uneven development is a fundamental mechanism of the capitalist mode of production (Amin, 1974; Aglietta, 1979; Harvey, 1982). It refers to the differential rate of investment (and, therefore, of allocation of resources) between economic sectors and spatial areas, according to their changing potential of profitability. It follows a differential rate of growth that, by the different level of wealth it generates in various places and activities, tends to have cumulative effects: Areas that first rose to economic and social prominence have an advantageous edge over the laggard ones as far, and only as far, as there is no major altering of the historical pattern of economic growth. When such an alteration occurs, we witness a transformation of the spatial structure, a process that is known as *regional restructuring* (Markusen, 1984a; Friedman and Weaver, 1979). The entire world is currently undergoing such a process, in an attempt to supersede the structural crisis of capitalism by continuously shifting location of capital, labor, markets, production, distribution, and management to the most advantageous location for each economic unit. High technology plays a major role in such a process, both as a technological tool and as the core of new sectors of economic growth.

The spatial effects of this new form of uneven development are being felt both between countries and areas of the world and between regions of each country. I will examine the effects of these two processes (international and interregional) *on the spatial structure of the United States*, leaving aside, because of obvious constraints on the length of this text, the overall spatial change at the international level.

In recent years the debate on regional development has been dominated by the "sunbelt/snowbelt" problematic (Perry and Watkins, 1977; Sale, 1975; Sternlieb and Hughes, 1976; Weinstein and Firestine, 1978). At first sight, investment, jobs, people, and income are leaving in growing numbers the traditional Northeastern and Northcentral centers of American capitalism to start a new process of growth in the more promising lands of the Southwest, of the West, and, to some extent, of the South (particularly in Florida) (Business Week, 1981:152). According to a standard interpretation, the movement to the sunbelt is motivated by a search by business for lower wages, nonunionized labor, and less taxes and government control. Nevertheless, the counter-example of New England (clearly a part of the new dynamic pole in spite of its union tradition), the significant level of government intervention in California, and the new economic dynamism of New York point toward the need for a more subtle analysis of the new regional dynamics of the United States. Mollenkopf (1983) has suggested that the main explanatory variable is the intersectorial differentiation between old line manufacturing and new industries and services. Those cities and regions that concentrate traditional sectors (such as Detroit or Cleveland) experience a net decline; those that start their develop- ment with the new activities (such as Denver or San Jose) experience the highest rates of growth; and those like Boston or New York, in which old and new sectors coexist, undergo simultaneous growth and decline. Also, Bluestone and Harrison give a more precise meaning to the "deindustrialization thesis" (Bluestone and Harrison, 1982) that they have to popularize, when they write the following:

> In contrast to others (e.g., Schultze and Lawrence), who choose to characterize it as mainly an aggregate phenomenon, the deindustrialization of America refers to a process of unequal and uneven development, according to which particular sectors, regions, and groups of citizens play a disproportionate price for the new competitive strategies being pursued by American business. We have seen this to be clearly the case for different industries, races, genders, (and regions) [1984:20].

So it is the new wave of uneven development, with its spearheads mainly in advanced services and in high technology manufacturing, that appears to be the driving force underlying the current regional restructuring (Sawers and Tabb, 1984).

The new spatial division of labor in the United States also results from the growing internationalization of the economy, accelerated after the 1974-1975 crisis on several levels (Glickman, 1980). Take, for example, the increasing role of international trade, with the United States importing more and more manufactured goods while exporting services, agricultural products, and high tech manufacturing. And capital itself has become more internationalized. Now, along with the growing outflow of U.S. capital abroad, there has been a massive increase of direct foreign investment in the United States, which has grown 600% between 1973 and 1983 until reaching $111.3 billion (Business Week, 1983). Much of this foreign investment is in manufacturing, in both old and new sectors (Schoenberger, 1985).

A third trend concerns the renewed internationalization of labor in the United States (Glickman and Petras, 1981), in which we witness an immigration flow unprecedented since the 1920s, fueling, among other things, the informal economy in the largest cities (Portes, 1985; Sassen-Koob, 1984a). There is also an internationalization of the production process, with location of different facilities in different countries, and reintegration of the process through improved communications (Mutlu, 1979; Henderson and Scott, 1984). Along with electronics, the automobile industry is the most significant example of this "global factory" with the production of the "world car," whose different components are produced in various countries and assembled on the final markets: By 1980 37.2% of the total motor vehicle production of the four leading U.S. automobile companies were located abroad (Trachte and Ross, 1983). Another expression of the new internationalization of production is the formation of the so-called "border regions," particularly in the United States. Mexico's Border Region has profoundly transformed the socioeconomic structure of large areas in Mexico and in the United States as well (Hansen, 1981).

Finally, the process of internationalization in the United States has a specific historical actor: the multinational corporation. It is the emergence of this particular type of organization—transcending national boundaries, cultural specificity, and political controls within a global strategy, whose logic cannot be fully understood in any given spatial unit—that allows for the expression of the four processes of internationalization we have described (United Nations, 1978; Palloix, 1977; Vernon, 1971; Barnett and Muller, 1974). So we find a tendency both toward the delocalization of the logic of the economic process and toward the concentration of decision-making units in a few commanding heights of the international economy.

High technology plays a major role in this process of internationalization on different levels. First, it allows communication and decentralized unified management between spatially scattered units, through the new telecommunication technologies. Second, high technology manufacturing epitomizes the new spatial division of labor, with the locationally distinct hierarchy between research and design and assembly-line operations, therefore spearheading the new space of global production, facilitated by the light weight of many electronic components, whose value is basically due to their informational content. Third, the process of automation and the increasing precision of machines (through robots and numerical machine tools) make possible a large-scale standardization of the components of most manufacturing activities and their recombination wherever the location appears to be convenient.

Finally, we should remember that high technology is, above all, a new process of production and management. Therefore, the more the economy becomes open, internationalized, and competitive, the more the appropriation of high tech and its implementation in the factories, offices, and communication systems of a company or of a country become crucial elements in winning a competitive edge. In this sense, the internationalization of the economy greatly reinforces the growth and importance of high tech activities. Because of this importance, investment in high tech production and the conquest of markets for high tech products in growing demand become key elements in the strategy to dominate the international economy (Zysman and Cohen, 1983).

We are now in a position to summarize the current trends of spatial restructuring in the United States in the context of the new international economy and in interaction with high technology activities:

(1) The internationalization of the economy actually reinforces the spatial polarization between sectors in different regions and within their metropolitan areas. Particularly because manufacturing exports are decreasing and manufacturing imports are on the rise, the old line industries are increasingly hurt by international competition. Furthermore, outflow of capital investment (for instance, in the automobile) and shift of capital toward the promising high tech sectors will augment intersectoral differentiation and, therefore, the distance between spatial areas we have been able to associate with each economic sector.

(2) The concentration of economic power on a world level in a few hundred major corporations is spanning growth of advanced corporate services and headquarters in a few major metropolises, consolidating the formation of what Friedmann and Wolff have called *the world city* (Friedmann and Wolff, 1982).

(3) The new process of international migration concentrates a new labor force and, therefore, the informal economy precisely in these world cities, whose "underground" component also has to be underlined as a part of the same system.

(4) Finally, perhaps the most striking effect of the new international economy on cities and regions is the loss of their autonomy vis-à-vis the worldwide economic actors that control their activities in terms of a global logic largely ignored and uncontrolled by local and regional societies. A rapidly changing economic space determined by economic units whose size and transnationality places them above social pressures and political controls is a *tendency* that, favored by the internationalization process *and* by high tech, attempts to impose the abstraction of a space of strategic decisions over the experience of place-based activities, cultures, and politics. Yet the reactions of *developmental states* (such as France, Japan, or Brasil) in the international economy creates a new dynamics, equally abstract in its horizon, but more directly rooted in political pressures and social values (Zysman and Cohen, 1983). High technology and the process of internationalization give rise to the space of flows, but the political dynamics of the decision-making process in the world economy partially restores the space of historical meaning. It is in the middle of this dialectics that cities and regions live, die, struggle, change. And with them, people.

CONCLUSION

The spatial structure of the United States has undergone a fundamental change in the last decade, as is shown, somewhat schematically, by the results of the 1980 U.S. Census (Long and De Are, 1983). It is my hypothesis that a major underlying cause of this change has been the interconnected processes of economic restructuring and technological change examined here. If this is true, we are only at the beginning of this process and the forces at work are likely to modify in a few years the entire urban problematic (Salomon, 1980; Bradbury et al., 1982; Noyelle and Stanback, 1984).

Therefore, a useful way to conclude this chapter will be to relate the basic trends of the transformation of the U.S. spatial structure in the 1970-1980 period to the processes analyzed here. The whole structure appears to follow a logic that combines *spatial diffusion, territorial hierarchy (including an urban concentration of the highest level)*, and *functional interconnectedness*.

If we refer these tendencies (that I am unable to relate here) to the information and ideas presented in this chapter, it is easy to observe the close connection between the urban-regional consequences of the processes of economic-technological restructuring and the emerging spatial forms: First, high tech-led economic development and the search for less astringent conditions for capital have fostered the new interregional division of labor. Second, advanced corporate services required by the informational mode of development and by the global needs of multinational capitals have provoked the revitalization of high-level CBDs and created the conditions for a limited, yet significant, gentrification of their adjacent neighborhoods. Third, the transition from the Welfare State to the Warfare State, along with the planned shrinkage of urban services, has precipitated both the decline of large central cities and the expansion of suburbs, particularly in the regions where the "new economy" is booming. Fourth, the interpenetration of economies and societies at the international level, facilitated by the new communication technologies, has laid the ground for the "immigrant city," as well as for the ex-industrial "ghost towns" that often become the "back door" for the newcomers into the city. Fifth, the polarized occupational structure favored by high tech, both in its production and in its applications, has considerably contributed to the deepening of cleavages between cities, between inner cities and suburbs, and within central cities. And, finally, new communication technologies, the coming of the "electronic home," and the "electronic office" have stimulated territorial sprawl, suburbanization, and individualization of sociospatial relations.

Altogether, two major processes appear to express, at the spatial level, the dynamics of the current technological and economic restructuring: (1) New technologies allow the emergence of a *space of flows,* substituting for a space of places, whose meaning is largely determined by their position in a network of exchanges. The logic of large-scale organizations fits perfectly into a spatial form that abstracts from historical reality and cultural specificity to accommodate new information and instructions. And (2) the new model of capitalist growth, supported by the Warfare State and the informational mode of production, induces a *new territorial division of labor,* based on *polarized growth* and *selective development,* whith reflects itself in the interregional cleavages, intrametropolitan dualism and *simultaneous* life and death of our great cities.

Thus, a pattern of coherent relationships appears to exist between the processes I have analyzed and the recent evolution of the U.S. spatial structure. Since 1980 these trends have clearly accelerated in similar directions, so it seems that we are only observing the early stages of a new urban-regional process.

To be sure, other factors account for the evolution of spatial forms: demographic trends, such as the rise and fall of the "baby boom" and the aging of the population (Alonso, 1983); cultural patterns, such as revolution in the family and use-value orientation of the new urban culture (Hayden, 1984); and sociopolitical processes such as the new neighborhood movement, the rejuvenation of local politics, the emergence of women and minorities as political forces, and the fundamental clash over criteria defining the "quality of life" (Castells, 1983b). In this sense, the analysis of the transformation of the spatial structure must include both the technoeconomic restructuring of our societies and the social, political, and cultural debates trying to shape, alter, or even reverse the process currently under way. Yet unless these debates place themselves within the actual realm of technoeconomic transformation that I have tried to identify in this chapter, urban darkness will be likely to overshadow city lights.

NOTES

1. This chapter attempts to synthesize the main substantive tendencies observed in a wide range of fields concerning the interaction between high technology and spatial structure in the United States. In order to provide the essence of the

observations, as well as some analytical hypotheses, little room has been left for a presentation of the empirical data themselves. The interested reader is referred to the sources listed in the reference section of this chapter. A detailed discussion of the research findings underlying the argument presented in this chapter can be found in Castells (1984).

2. There is some confusion regarding the actual pattern of location of high technology activities, depending on the sectors that are included in the analysis. Nevertheless, the most complete empirical study (Glasmeier et al., 1983), while reporting the heterogeneity of the locational patterns observed, does show some interesting tendencies in terms of the metropolitan areas that attract most high tech employment. For the period studied (1972-1977, corresponding to the last two U.S. censuses of manufacturing available in 1983), the top two winners in high tech jobs were San Jose and Anaheim (CA). These cities were followed at some distance by Houston, San Diego, Boston, Dallas, Worchester (MA), Oklahoma City, Lakeland (FL), and Phoenix. New York, Philadelphia, Cleveland, Miami, and Syracuse were the top losers. Concerning high tech plants, the winning list should also include Los Angeles, Chicago, Minneapolis, San Francisco, and Detroit. New York continues to be the biggest loser in this area.

Storper (1982) reported that for 1980 the top four states employing people to work on semiconductors (SIC 3674 and part of SIC 3679) account for 72.5% of all such jobs in the nation. California alone accounts for 30.4%, followed by Arizona (18.3%), Texas (14.6%), and Massachusetts (9.2%). These data fit well with the traditional image of high tech areas, but Storper also shows that in terms of value-added, in 1977, the top states were California (34.4%), Texas (17.2%), *New York (12.3%), and Pennsylvania (10.5%)*. Glasmeier (1985) also reports that the old industrial states account for a substantial proportion of high tech employment and plants, particularly if one takes a broad definition of high tech manufacturing, including 100 sectors.

Indeed, it seems that we must differentiate between the existing situation (and even more, the situation of the 1970s) and the current trends of high tech location. A study by Pinkerton (1984) ranks the states in terms of the absolute values of the two key high tech sectors in 1975-1981, and then measures the trends of change between the two sets of data. While the old industrial states still do quite well in absolute terms, only New England joins the West and the Southwest in the new developments. Because I am interested more in the logic of the new space of production than in the mapping of high tech employment at one point in time, I retain the trends of high tech location for the 1980s as representative of the spatial characteristics associated with the new industry.

3. Unfortunately, there is very little reliable evidence on the interaction between communication technologies and social relationships in the shaping of the city. Therefore, my analysis on this issue will be tentative and exploratory, based primarily on hypotheses on the potential effects of new technologies, given the current knowledge and experience on urban social relations. Yet the issue is so important that it seems worthwhile to take the risk of advancing some ideas even before they are transformed, as is my intention, in scholarly research.

4. Nevertheless, it would be erroneous to conclude that high technology is an instrument of capital. New technologies could allow, for instance, reducing working time while still improving our standards of living and broadening the realm of our experience. It is our social organization and subordination of technology to business's strategies that transforms into hardship (fewer jobs, lower wages) what could be a tool for human liberation. Technologies supress working time, not jobs.

5. For example, Nobel Laureate Wassily Leontieff declared:

The computer and the robot are already beginning to replace the simple mental functions of blue and white collar workers. Man, as a factor of production, has only two aspects: physical and mental. Both are being replaced. The only solution is to stretch out vacations, shorter work hours, and share available jobs (Time, 1983).

REFERENCES

ADAMS, P. and G. FREEMAN (1982) "Social services under Reagan and Thatcher," pp. 65-82 in M. I. Fainstein and S. S. Fainstein (eds.) Urban Policy Under Capitalism. Beverly Hills, CA: Sage.

AGLIETTA, M. (1979) A Theory of Capitalist Regulation: The U.S. Experience. London: New Left Books.

ALONSO, W. (1983) "The demographic factor in housing for the balance of this century," pp. 33-50 in G. W. Gau and M. A. Goldberg (eds.) North American Housing Markets into the Twenty-First Century. Cambridge, MA: Ballinger.

AMIN, S. (1974) Le Developpement Inegal. Paris: Editions de Minuit.

BALDWIN, T. and D. S. McVOY (1983) Cable Communication. Englewood Cliffs, NJ: Prentice-Hall.

BALDWIN, T., J. ABEL, and E. SKINNER (1980) "New communication technologies and mass media environment: a question of access." National Forum (summer): 28-30.

BARAN, B. (1984) Insurance: Draft Research Report. Berkeley: University of California, Berkeley Roundtable on the International Economy.

———(1982) "The transformation of the office industry: impacts on the workforce." Masters thesis, University of California, Berkeley.

———and S. TEEGARDEN (1983) Women's Labor in the Office of the Future: Research Paper. Berkeley: University of California, Institute of Urban and Regional Development.

BARNETT, R. and R. MULLER (1974) Global Research. New York: Simon & Schuster.

BLUESTONE, B. and B. HARRISON (1984) "The economic state of the union in 1984: uneven recovery, uncertain future." (unpublished)

———(1982) The Deindustrialization of America. New York: Basic Books.

BORRUS, M. and J. MILLSTEIN (1982) Technological Innovations and Industrial Growth: A Comparative Assessment of Biotechnology and Semiconductors: Research Report Prepared for the U.S. Congress Office of Technology Assessment. Berkeley: University of California, Berkeley Roundtable on the International Economy.

BOWLES, S., D. GORDON, and T. WEISSKOPF (1983) Beyond the Wasteland. New York: Doubleday.

BRADBURY, K., A. DOWNS, and K. SMALL (1982) Urban Decline and the Future of American Cities. Washington, DC: Brookings Institute.

Business Week (1984a) January 23: 99.

———(1984b) "The consumer rush is on for anything electronic." February 27: 148-150.

————(1983) June 4: 103-104.

————(1981) "America's restructured economy." June 1: entire issue.

CALHOUN, C.J. (1981) "The microcomputer revolution? Technical possibilities and social choices." Sociological Methods and Research (May): 415.

CARLSON, R. and T. LYMAN (1984) U.S. Government Programs and Their Influence on Silicon Valley: Research Report. Menlo Park, CA: SRI International.

CARNOY, M. (1984) The State and Political Theory. Princeton, NJ: Princeton University Press.

————and M. CASTELLS (1984) "After the crisis?" World Policy Journal (Spring): 495-516.

CARNOY, M. and D. SHEARER (1980) Economic Democracy. New York: Sharpe.

————and R. RUMBERGER (1983) A New Social Contract. New York: Harper & Row.

CASTELLS, M. (1984) Towards the Informational City? High Technology, Economic Change, and Spatial Structure: Some Exploratory Hypotheses; Working Paper 430. Berkeley: University of California, Institute of Urban and Regional Development.

————(1983a) "Crisis, planning, and the quality of life." Society and Space 1: 1.

————(1983b) The City and the Grassroots. Berkeley: University of California Press.

————(1980) The Economic Crisis and American Society. Princeton, NJ: Princeton University Press.

————(1979) "The service economy and the post-industrial society." International Journal of Health Services 6(4).

CHANDLER A. D. (1977) The Visible Hand. Cambridge: Cambridge University Press.

CLAYTON, J.L. (1962) "Defense spending: key to California's growth." Western Political Quarterly 15(2).

COHEN, R. (1978) The Corporation and the City. New York: Columbia University, Conservation of Human Resources Project.

CROUCH, C. [ed.] (1979) State and Economy in Contemporary Capitalism. London: Croom Helm.

DANIELS, P. W. [ed.] (1979) Spatial Patterns of Office Growth and Location. New York: John Wiley.

DE GRASSE, R. (1983) Military Expansion, Economic Decline: The Impact of Military Expending on U.S. Economic Performance. New York: M. E. Sharpe.

DELLUMS, R. [ed.] (1983) Defense Sense: The Search for a National Military Policy. Cambridge, MA: Ballinger.

DEMPSEY, R. and D. SCHURDE (1971) "Occupational impact of defense expenditures." Monthly Labor Review (December): 12-15.

DENISON, E. (1979) Accounting for Slower Economic Growth. Washington, DC: Brookings Institution.

DERTOUZOS, M. L. and J. MOSES [eds.] (1981) The Computer Age: A Twenty Year View. Cambridge, MA: MIT Press.

DORDICK H.S., H. G. BRADLEY, B. NAMES, and T. H. MARTIN (1979) "Network information services: the emergence of an industry." Telecommunication (September): 217-234.

DUMAS, L. [ed.] (1982) The Political Economy of Arms Reduction. Boulder, CO: Westview.

EDP Analyzer (1982) "Experiences with tele-commuting." 20(November): 11.

FRIEDMANN, J. and C. WEAVER (1979) Territory and Function. London: Edward Arnold.

FRIEDMANN, J. and G. WOLFF (1982) World City Formation. Los Angeles: University of California, Comparative Urbanization Studies.

FROBEL, F. et al. (1980) The New International Division of Labor. Cambridge: Cambridge University Press.

GANSLER, J. (1980) The Defense Industry. Cambridge, MA: MIT Press.

GLASMEIER, A. (1985) "Spatial differentiation of high technology industries: implications for planning." Ph.D. dissertation, University of California, Berkeley.

——P. HALL, and A. MARKUSEN (1984) "Can everyone have a slice of the high tech pie?" Berkeley Planning Journal 1(summer).

——(1983) Recent Evidence on High Technology Industries' Spatial Tendencies: Working Paper. Berkeley: University of California, Institute of Urban and Regional Development.

GLICKMAN, N.J. (1983) "Economic policy and the cities: in search of Reagan's real urban policy." Presented at the North American Meeting of the Regional Science Association, Chicago.

——(1980) "International trade, capital mobility, and economic growth: some implications for American cities and regions." Reported to the President's Commission for a National Agenda for the 1980s, Washington, DC.

——and E.M. PETRAS (1981) International Capital and International Labor Flows: Implications for Public Policy; Working Papers in Regional Science. Philadelphia: University of Philadelphia.

GORDON, D., R. EDWARDS, and M. REICH (1982) Segemented Work, Divided Workers. Cambridge: Cambridge University Press.

GOUGH, I. (1979) The Political Economy of the Welfare State. London: Macmillan.

HALL, P., A. MARKUSEN, R. OSBORN, and B. WACHSMAN (1983) The Computer Software Industry: Prospects and Policy Issues; Working Paper 410. Berkeley: University of California, Institute of Urban and Regional Development.

HANSEN, N. (1981) "Mexico's border industry and the international division of labor." Annals of Regional Science 15(1).

HARVEY, D. (1982) The Limits of Capital. Oxford: Blackwell.

HAYDEN, D. (1984) Redesigning the American Dream. New York: Norton.

HENDERSON, J. and A. SCOTT (1984) The American Semiconductors Industry and the New International Division of Labor: Working Paper. Hong Kong: University of Hong Kong, Center of Urban Studies and Planning.

HENRY, D. (1983) "Defense spending: a growth market for industry," in U.S. Industrial Outlook 1983. Washington, DC: U.S. Government Printing Office.

HICKS, D.A. and N.J. GLICKMAN [eds.] (1983) Transition to the 21st Century: Prospects and Policies for Economic and Urban-Regional Transformation. Greenwich, CT: JAI Press.

HIRSCHHORN, L. (1984) Beyond Mechanization. Cambridge, MA: MIT Press.

——et al. (1983) Cutting Back: Retrenchment and Redevelopment in Human and Community Services. San Francisco: Jossey Bass.

JANOWITZ, M. (1980) The Last Half Century. Chicago: University of Chicago Press.

LECKACHMAN, R. (1981) Greed Is Not Enough: Reaganomics. New York: Pantheon.

LEONTIEFF, W. and F. DUCHIN (1983) Military Spending. New York: Oxford University Press.

LONG, L. and D. DE ARE (1983) "The slowing of urbanization in the U.S." Scientific American 249(1): 33-39.

MAHONEY, S., N. DEMARTINO, and R. STENGEL (1980) Keeping Pace with the New Television: Public Television and Changing Technology. New York: Carnegie Corp., VNY Books.

MARK, J. A. and W. H. WALDORF (1983) "Multifactor productivity: a new BLS measure." Monthly Labor Review 106(12): 3-15.

MARKUSEN, A. (forthcoming) "Defense spending: a successful industrial policy?" International Journal of Urban and Regional Research.

———(1984a) Defense Spending and the Geography of High Tech Industries: Working Paper. Berkeley: University of California, Institute of Urban and Regional Development.

———(1984b) Profit Cycles, Oligopoly, and Regional Development. Cambridge, MA: MIT Press.

———(1983) High Tech Jobs, Markets, and Economic Development Prospects: Working Paper, Berkeley: University of California, Institute of Urban and Regional Development.

MARTIN, T. H. (1981) Telematic Society. Englewood Cliffs, NJ: Prentice-Hall.

MOLLENKOPF, J. (1984) "The post-industrial transformation of the political order in New York City." Presented at the Social Science Research Council's Conference on New York City, New York.

———(1983) The Contested City. Princeton, NJ: Princeton University Press.

———(1982) "Economic development," in C. Brechner and R. D. Horton (eds.) Setting Municipal Priorities. New York: New York University Press.

MOORFOOT, R. (1982) Television in the Eighties: The total Equation. London: British Broadcasting Co.

MUTLU, S. (1979) "Inter-regional and international mobility of industrial capital: the case of the American automobile and electronics industry." Ph.D. dissertation, University of California, Berkeley.

Newsweek (1984) "The video revolution." August 6: 32-39.

New York Times (1984) "Cable operators take a bruising." March 4.

NICOL, L. (1983a) Communications, Economic Development, and Spatial Structures: A Review of Research; Working Paper. Berkeley: University of California, Institute of Urban and Regional Development.

———(1983b) Information Technology, Information Networks, and On-Line Information Services: Working Paper. Berkeley: University of California, Institute of Urban and Regional Development.

NORTON, J. (1983) "Keeping up with households." American Demographic Journal (February): 17-21.

NOYELLE, T. and T. STANBACK (1984) The Economic Transformation of American Cities. Totowa, NJ: Rownan and Allanheld.

O'CONNOR, J. (1984) Accumulation Crisis. Oxford: Blackwell.

Organization for Economic Cooperation and Development (1981) The Welfare State in Crisis. Paris: Author.

PALLOIX, C. (1977) L'economic modiale capitaliste et les firmes multinationales. Paris: Maspero.

PALMER, J. A. and I. SAWHILL [eds.] (1982) The Reagan Experiment. Washington, DC: Urban Institute Press.

PERRY, D. C. and A. J. WATKINS (1977) The Rise of Sunbelt Cities. Beverly Hills, CA: Sage.

PINKERTON, S. J. (1984) High Technology Growth and Regional Structure: Seminar Paper for PLUS 508. Los Angeles: University of Southern California, School of Urban Planning.

PORTES, A. (1985) Latin Journey, Cuban and Mexican Immigrants in the U.S. Berkeley: University of California Press.

———and J. WALTON (1981) Labor, Class, and the International System. New York: Academic.

RICE, R. E. and Associates (1984) The New Media. Beverly Hills, CA: Sage.

ROGERS, E. M. and J. K. LARSEN (1984) Silicon Valley Fever Growth of High Technology Culture. New York: Basic Books.

RUTZIEK, M. A. (1970) "Skills and location of defense-related workers." Monthly Labor Review 93: 11-16.

SABEL, C. (1982) Work and Politics: The Division of Labor in Industry. Cambridge: Cambridge University Press.

SALE, K. (1975) Power Shift. New York: Random House.

SALOMON, A. P. [ed.] (1980) The Prospective City. Cambridge, MA: MIT Press.

SASSEN-KOOB, S. (1984a) "The new labor demand in global cities," pp. 139-172 in M. P. Smith (ed.) Cities in Transformation. Beverly Hills, CA: Sage.

———(1984b) "Growth and informalization at the core: the case of New York City." Presented at the Seminar on the Urban Informal Economy in Core and Periphery, Johns Hopkins University, Department of Sociology, Baltimore.

SAWERS, L. and W. TABB [eds.] (1984) Sunbelt/Snowbelt, Urban Development, and Regional Restructuring. New York: Oxford University Press.

SAXENIAN, A. L. (1984) "Silicon Valley and Route 128: regional prototypes or historic exceptions." Presented at a Conference on Microelecronics, University of California, Santa Cruz.

———(1980) "Silicon chips and spatial structure: the industrial basis of urbanization in Santa Clara County, California." Master Thesis, University of California, Berkeley.

SCHEMENT, J. R., L. A. LIEVROW, and H. S. DORDICK (1983) "The information society in California: social factors influencing its emergence." Telecommunications Policy 7(1): 64-72

SCHOENBERGER, E. (1985) "The regional impact of foreign investment in U.S. manufacturing." Ph. D. dissertation, University of California, Berkeley.

SERRIN, W. (1983) "High tech is no jobs panacea, experts say." New York Times (September 18).

SINGELMANN, J. (1977) The Transformation of Industry from Agriculture to Service Employment. Beverly Hills, CA: Sage.

SOJA, E., R. MORALES, and G. WOLFF (1983) "Urban restructuring: an analysis of social and spatial changes in Los Angeles." Economic Geography 59(2): 195-230.

STANBACK, T. (1979) Understanding the Service Economy. Baltimore: Johns Hopkins University Press.

———P. J. BEARSE, T. J. NOYELLE, and R. A. KARASEK (1981) Services: The New Economy. Totowa, NJ: Allanheld and Osmun.

STARK, P. (1984) "Special report: VCR revolution, the big changes in entertainment." San Francisco Chronicle (February 27).

STERNLIEB, G. and J. W. HUGHES (1976) Postindustrial America: Metropolitan Decline and Inter-Regional Job Shifts. New Brunswick, NJ: Rutgers University, Center for Urban Policy Research.

STORPER, M. (1982) "The spatial division of labor: technology, the labor process, and the location of industries." Ph.D. dissertation, University of California, Berkeley.

STRASSMAN, P. (1980) "The office of the future: information management for the new age." Technology Review (December).

SYLVESTER, E.J. and L.C. KLOTZ (1983) The Gene Age: Genetic Engineering and the Next Industrial Revolution. New York: Scribner.

TEITZ, M. (1984) "The California economy, changing structure and policy responses," in J.J. Kirlin and D.R. Winkler (eds.) California Policy Choices, 1984. Los Angeles: University of Southern California, School of Public Administration.

THUROW, L.C. (1984) "The disappearance of the middle class." New York Times (February 5).

Time (1983) Interview with Wassily Leontieff. May 30.

TOMASKOVIC-DEVEY, D. and S.M. MILLER (1982) "Recapitalization: the basic U.S. urban policy of the 1980s," pp. 23-42 in N.I. Fainstein and S.S. Fainstein (eds.) Urban Policy Under Capitalism. Beverly Hills, CA: Sage.

TRACHTE, K. and R. ROSS (1983) "The crisis of Detroit and the emergence of global capitalism." Presented at the Meeting of the American Sociological Association, Detroit.

United Nations (1978) Transnational Corporations in World Development: A Reexamination. New York: Author.

VERNON, R. (1971) Sovereignty At Bay: The Multinational Spread of U.S. Enterprise. New York: Basic Books.

WALKER, P. (1983) "The distribution of skill and the division of labor, 1950-78." Ph.D. dissertation, University of Massachusetts.

WALKER, R. and M. STORPER (1984) "The spatial division of labor: labor and the location of industry," p. 19-47 in L. Sawers and W. Tabb (eds.) Sunbelt/Snowbelt, Urban Development and Regional Restructuring. New York: Oxford University Press.

WAYNE, L. (1984) "America's astounding job machine." New York Times (June 17).

WEBBER, M. (1980) A Communications Strategy for the Cities of the 21st Century: Working Paper. Berkeley: University of California, Institute of Urban and Regional Development.

WEINSTEIN, B.L. and R. FIRESTINE (1978) Regional Growth and Decline in the U.S. New York: Praeger.

WILLIAMS, F. (1982) The Communications Revolution. Beverly Hills, CA: Sage.

WILENSKY, H. (1974) The Welfare State and Equality. Berkeley: University of California Press.

WILMOTH, D. (1983) "The evolution of national urban policy in the U.S." Ph.D. dissertation, University of California, Berkeley.

WOLFE, A. (1976) The Limits of Legitimacy. New York: Pantheon.

ZYSMAN, J. and S. COHEN (1983) The Mercantilist Challenge to the Liberal International Trade Order. Washington, DC: U.S. Congress, Joint Economic Committee.

2

Technology, Space, and Society in Contemporary Britain

PETER HALL

☐ EARLIER and more severely than most nations, Great Britain has suffered from a problem of deindustrialization (Blackaby, 1979), the causes and spatial repercussions of which have been intensively analyzed by economists and geographers (Fothergill and Gudgin, 1982; Massey and Meegan, 1982; Massey, 1983). Appropriately, both academics and politicians were early to grasp the relevance of the newer high technology "sunrise" industries as a route to economic regeneration.

At the political level, even before 1979 the Callaghan Labour government was instrumental in the creation of INMOS, an indigenous company for silicon chip manufacture, and the Thatcher government followed the same line in designating 1982 Information Technology Year (Baker, 1982). At the level of academic research, British universities have led in analysis; first, of technological progress and economic development at the Science Policy Research Unit of the University of Sussex (Rothwell and Zegveld, 1981; Freeman et al., 1982; Freeman, 1984); and second, of the regional implications of technological innovation and its diffusion at the Centre for Urban and Regional Development Studies of the University of Newcastle upon Tyne (Oakey et al., 1980; Thwaites, 1982; Oakey, 1983).

TWO GEOGRAPHIES: DEINDUSTRIALIZATION, REINDUSTRIALIZATION

That research has thrown strong light on the phenomenon, but it has also illuminated some uncomfortable features of the new geography of innovation—and it has left in doubt the precise mix of policies that might help ameliorate them. The overwhelming fact, underlined in all the research so far available, is that the new high technology industries are growing up in regions and in places very different from those in which the older industries are declining.

Britain's declining basic industries—coal, shipbuilding, heavy engineering, textiles—are strongly concentrated in the regions in which they were originally established in the nineteenth century: Central Scotland around Glasgow, Northeast England around Newcastle upon Tyne and Sunderland, the Northwest around Liverpool and Manchester, and South Wales focused on Swansea and Cardiff. Interventionist regional policies had a fair measure of success during the boom years of the 1960s in steering newer industries (such as cars and electrical engineering) into these regions. In the great recession of the late 1970s and early 1980s, however, new and disturbing features have emerged. The group of declining industries has now extended to include many—particularly the car industry— previously thought immune, and similarly the spatial extent of industrial decline has broadened to include major urban areas such as Greater London and the West Midlands metropolitan county, which formerly were counted as part of prosperous Britain.

The new high technology growth, in contrast, has been limited to a few areas outside these major industrial regions, notably, the belt along the M4 motorway from London to Bristol and the region around Cambridge, which Britain's regional policy makers, located in the Department of Trade and Industry, have traditionally considered to be zones of restraint. This conclusion, it needs stressing, is based on direct observation, surmise, and some minimal statistical observation. The basic statistical data gathering about the growth of high technology industry in Britain is still under way and is not yet ready to be reported.

THE LOCATION OF R&D

Perhaps the most important information so far available concerns, first, the geography of research and development and, second, the industrial take-up of innovations by region. The recent work by Howells (1984) confirms the much older study by Buswell and Lewis (1970): Research and development services in Britain are disporportionately concentrated in the Southeast (with over 57% of total British employment in R&D in 1976) and have been disproportionately growing there, with a 34% growth between 1971 and 1976 against a national expansion of less than 22 % (Howells, 1984: 16). But Howells also shows that this concentration is outside Greater London in the more rural counties of the Southeast, especially west of London, in the counties alongside the so-called M4 corridor. He further concludes that this movement to more rural areas has been led by the large research facilities of multisite companies, requiring good contacts between their laboratories and their headquarter offices, which are themselves tending to disperse outward from London to the outer southeast (Howells, 1984: 26).

The work by the CURDS group provides an interesting complement to Howells's conclusions. Oakey et al. showed in 1980 that innovative manufacturing establishments were highly concentrated in the East Anglia, Southwest, and Southeast regions; the least innovative regions, in their analysis, were the older peripheral industrial areas of the Northwest, Scotland, and Wales (Oakey et al., 1980: 243). The number of innovating plants in the Southeast was particularly high when the plant structure was taken into account (Oakley et al., 1980: 248). Later CURDS work on product innovation followed up the effect of plant ownership, concluding that in the peripheral development areas much of the innovation was in branch firms controlled from outside, and that the level of innovation in indigenously-owned plants was notably low (Thwaites, 1982: 375). All this may suggest that the concentration of R&D facilities in the Southeast may be uniquely favorable to a high level of innovation there but that, because this concentration tends to result from the concentration of headquarter offices there, it may be difficult to shake.

This pattern, it perhaps needs stressing, appears to be very different from the American pattern. In the United States, Malecki's work has shown that R&D tends to be concentrated in a number of different locations—some of them representing the old established headquarters of manufacturing firms in the Northeast and Midwest, some of them new, innovatory, federally-supported research centers in the Sunbelt (Malecki, 1980: 229-231; Malecki, 1981a: 128-130; Malecki, 1981b: 80). The differences seem to arise for a number of reasons. First, large American corporations tend to have retained their headquarter offices next to their principal production facilities in the great manufacturing cities, rather than transferring them to a financial center such as New York; the dominance of the great corporate metropolis, so clear in Britain, is not so pronounced in the United States. Second, in the United States much Federal R&D has gone to Sunbelt locations in which universities or major research laboratories are present, whereas in Britain much of the equivalent support has gone to government research establishments on or close to the M4 corridor west of London. These differences in turn reflect the continental scale of the American economy and its relatively decentralized organizational structures, in contrast to the short distances and the London dominance that characterize the British space-economy.

THE HIGH TECHNOLOGY PHENOMENON: DEINDUSTRIALIZATION REVISITED?

As already stressed, all this evidence is still extremely fragmented. We are not clear about some of the most basic facts. In particular, there is a doubt as to perhaps the most basic fact of all: whether high technology growth actually represents the road to reindustrialization, or whether it merely represents another form of deindustrialization.

On this, the available evidence is contradictory. For the United States, Saxenian (1984) has shown that during the 1970s Silicon Valley gained an average of 25,000 new jobs a year, of which 10,000 were in manufacturing and the rest were in services; while Glasmeier (1983) suggest that in the entire United States between 1972 and 1977 high technology manufacturing contributed a net increase of some 378,000 new jobs, although some 37% of the individual high tech industries showed contraction in employment.

For Britain, recent and so far unpublished evidence shows that the position is (as is so often the case) gloomier (McQuaid, 1984). High tech industries (including R&D) accounted for 10.6% of all manufacturing employment and 5.2% of all employment in Great Britain in September 1981. On a slightly different definition (necessitated by a change in classification), of 9 high tech industries, only 2 (electronic computers; radio, radar, and electronic capital goods) showed any employment growth at all over the period 1974-1983, or over the shorter subperiod 1980-1983.

Evidence is not yet available for the geographical breakdown of these British figures. But it is significant that even in the area of Britain with the strongest high tech concentration—the Southeast region outside of London—manufacturing employment declined by 14.9% between 1974 and 1982, while service employment increased by 12% (Regional Studies Association, 1983: 10). Furthermore, in the very heart of the M4 corridor—the Central Berkshire area—manufacturing employment fell 10.6% between 1971 and 1977, whereas employment in business, professional services, and administration rose 29.4% (Breheny et al., 1984). True, both in the Southeast generally and in Central Berkshire there were big declines in older, nonhigh tech manufacturing during these periods; but, as Breheny et al. stress, the evidence strongly suggests that producer services, not manufacturing, provided the main engine of local economic growth during this period.

What is almost certainly happening is that high technology industry is following a general trend, whereby fewer and fewer people work in actual manufacturing processes while more and more work in service processes ancillary to, and necessary to, them (Gershuny and Miles, 1983: 65). A realistic economic accounting process would reclassify these latter workers to the relevant manufacturing industry. But in terms of occupational classification—and, therefore, of social class—the workforce ceases to be blue-collar, and becomes overwhelmingly white-collar, in character. To this point we shall return in the final section of this chapter.

THE NEW CORRIDORS OF HIGH TECH GROWTH

Given these facts, the emerging geography of high technology Britain is logical enough. It is almost certainly based on a very narrow group of information technology industries and on a wider range of associated producer services (on the nature of which, more research is going to be necessary). It stretches along corridors of high accessibility from London, marked by national intercity motorways (the M4 from London to Reading, Newbury, Swindon, and Bristol; the M1 from London to Leicester and Leeds; the M11 from London to Harlow and Cambridge) and by fast intercity trains. The main concentrations of high technology take the form of accretions to towns along these axes, some old (Cambridge, Reading), some new (Harlow, Milton Keynes, Bracknell, the expanding towns of Swindon and Basingstoke), generally within 80-100 miles from London at maximum, half that distance on average. The resulting spatial structure represents the progressive deconcentration of a dominant metropolitan city into a polycentric city region wherein some important controlling functions (finance, government) remain right in the urban core but others (including R&D, some headquarters, and specialized producer services) may be locally decentralized.

Ironically, this emerging spatial structure is precisely that which successive planning studies for the Southeast region during the 1960s and early 1970s sought to achieve. The (now defunct) South East

Economic Planning Council's Strategy, published in 1967, called for discontinuous sectors of growth along the main lines of communications outward from London (although the M11 was not identified as one) (Great Britain South East Economic Planning Council, 1967); the definitive Strategic Plan for the South East (1970) modified this so as to concentrate development into a limited number of major growth areas at varying distances from London, all of them, however, already identified as "study areas" in the earlier planning council strategy (Great Britain South East Joint Planning Team, 1970). It could be argued, therefore, that the planners' objectives have been realized. More likely, this has happened because the strategies were realistic in following strong trends that could even then be identified.

TOWARD EXPLANATION:
GENESIS OF THE M4 CORRIDOR

Beyond this outline, the research on the genesis of British high technology growth has yet to be completed. The central question is, of course, why the new areas of growth are in places quite different from the old areas of decline. One explanation is that these old areas suffer from an "Upas Tree" effect, whereby old, established industries and production methods exert a blighting effect on industrial innovation (Checkland, 1975; Hall, 1985: 10). As some evidence suggests that basic industrial innovations are bunched every half-century or so, thus triggering off so-called Kondratieff long waves in the capitalist economy, there is good reason that each successive innovation wave should occur in a new region (Hall, 1985: 10). Another theory is that both indigenous entrepreneurs and established multinationals are drawn to "amenity-rich locations," far from the blight of older industrial revolutions (Berry, 1970). Yet a third is that government policy has played a significant—and even a perverse—role by establishing government research centers around which new high technology industries incubate. Quite possibly, all three explanations may prove to have some force.

Here the stories of the M4 corridor and of the Cambridge area have obvious relevance. Unfortunately, although on both research is in progress, on neither have results been reported. In a preliminary paper, researchers at the University of Reading have set out some alternative hypotheses on the origins of the M4 corridor: first, that these lay in the location of government research establishments (GREs) during and immediately after World War II; second, that they lay in the decisions of American multinationals to locate close to Heathrow Airport's unique international connections, thus enjoying agglomeration economies together with other firms in the same fields (Breheny et al., 1985: 129). Possibly, they suggest in conclusion, both hypotheses are true: Originally developed by indigenous firms that had spun off from GREs, the corridor later attracted multinationals because of established agglomeration economies and good communications (Breheny et al., 1985: 129).

However, as already noted, these authors also draw a chastening empirical conclusion: High technology manufacturing was not the major element in the growth of the area during the 1970s when, indeed, manufacturing employment in Central Berkshire—the heart of the corridor—was in decline. Office development representing producer services, they suggest, was a far more significant force. The main reason was, in fact, technological: what they call a reduction in the specialized transportation cost of transmitting information, which frees offices from their earlier dependence on close face-to-face contact (Breheny et al., 1985: 125). It is the producer services that are decentralizing locally within the dominant metropolitan region but that still appear to be tied to it.

The Cambridge phenomenon is (so far) even more obscure. Available evidence suggests that the area has grown since 1960 principally through the foundation of new, often highly specialized, R&D consultancies in the field of CAD (Computer-Aided Design). A recent survey (Marsh, 1983) identified over 50 firms in the fields of computer hardware, software, electronics, instruments, lasers, and biotechnology, all founded since 1969, in the Cambridge area. Most remained small, although a few—including Sinclair Research and Acorn—had become British household names.

THE CONSEQUENCES FOR SOCIAL STRUCTURE

The results for social structure are quite profound. For the shopfloor jobs that are being lost in the older industrial locations are principally male, whereas the expanding service jobs are in traditionally female occupations, such as secretarial and clerical (Joseph, 1983: 195). Most of these female occupations are middle- to low-skill, middle- to low-income in character; during the entire period 1911-1971, and also in the subperiod from 1951 to 1971, women became increasingly overrepresented in the lower grades of work and underrepresented in the higher grades, in white-collar as well as blue-collar jobs (Hakim, 1979: 29). In this respect there seems to have been little change during the 1970s (Hakim, 1981: 524-525). Still, in the 1970s some 70% of women were working in a narrow range of occupations that were traditionally reserved almost exclusively to females, including typists, secretaries, maids, and nurses—the great majority of them service occupations in service industries (Joseph, 1983: 141).

It should not be concluded from this that women's earnings deteriorated; in fact, the earnings gap between women and men showed a steady narrowing over the long period from 1913 to 1971 for all groups except the semiskilled manual category (Joseph, 1983: 170). Joseph, who has critically examined the available economic theories of female job segregation, finds none without flaws and concludes that sociocultural factors may play an important role (Joseph, 1983: chap. 5, passim). The new female jobs, at any rate in the high technology industries and in the areas in which these concentrate, are almost certainly being filled by women whose husbands also work in white-collar occupations. The overwhelming character of the expanding settlements of the M4 corridor, and of similar areas, is therefore middle-class.

At the same time—as was also commented on by American observers for Silicon Valley—there is almost certainly a bipolar socioeconomic composition, with a relatively small number of very skilled and highly paid professionals and managers, still principally male, and a much larger number of lower-paid clerical ancillaries, principally female. What is not at all clear, pending detailed analysis of the social composition tables of the 1981 British Census, is how these two poles relate to each other within the same neighborhoods and even the same households. The conventional wisdom, well expressed in the influential book by Donnison and Soto (1980), is that the Good City—one might as well say the Good County or the Good Region—is the one in which the economy makes available middle-range, middle-paid jobs, thus giving lower-skilled people the opportunity to improve their (or the children's) positions. The information on which Donnison and Soto drew was derived principally from the 1960s, the period of the great boom in Britain's traditional basic industries, and is not directly relevant to an understanding of the new world of the M4 corridor.

POLICY PRESCRIPTION: BLIND NAVIGATION?

Given the almost complete lack of understanding of the geography of high technology—alike as to the basic facts of location, to the explanations, and to the consequences for socioeconomic structure—policy prescription would seem premature. Nevertheless, so attractive is the image of a high technology future that politicians and officials all over the country are inevitably drawn toward it. The most obvious expression in recent years is a veritable rash of Science Parks, more or less loosely modeled on Silicon Valley or North Carolina's Research Triangle, in every region; a recent survey counted more than 20 high technology development centers of different types (Worthington, 1982; Taylor, 1985). However, 8 among them were essentially commercial/business parks or upgraded industrial estates of a frankly commercial character, without an established research base; only 6 (Cambridge, Edinburgh, Aston, Warwick, Glasgow, and Salford) were true research parks set up on a university base. Taylor concludes, realistically, that "the commercial and municipal scramble to get a high-technology share at all costs" will inevitably lead to a high failure rate among ill-conceived or poorly sited developments, and that the forging of true academic-industrial links will prove vital (Taylor, 1985: 140-141).

The irony lies in the fact that the British government is vitally concerned with promoting the new technologies; yet the result, in Taylor's words, is "a totally uncoordinated picture, a locational scramble, a funding gamble" (Taylor, 1985: 142). This is because there is, as yet, no coordination between traditional regional policy—itself in ruins and subject to major review (Great Britain Department of Trade and Industry, 1983)—and educational-scientific policy. And, in turn, this almost certainly is because the latter is in the hands of a different government department (the Department of Education and Science, DES), which has proved remarkably indifferent to questions of regional or urban policy; to the extent that many of its key developments during the expansionist era of the 1960s, in new universities and polytechnics, flew in the face of regional priorities. By a supreme irony, DES has now succeeded in decentralizing the key decision units—the Research Councils, which fund much of the basic scientific research in universities and related institutions—to Swindon, in the heart of the M4 corridor.

SOME RESEARCH PRIORITIES

Better links between educational-scientific policy and regional-urban policy might help to remove such blatant contradictions. Whether they could prove successful in regenerating older blighted industrial regions is another matter; the Upas Tree effect might prove too strong after all. Critical, here, is the experience of those older regions that have had some modest success in the new technologies: Scotland's Silicon Glen in electronics, the Northwest region in R&D, America's Boston and Pittsburgh in computers and their applications. For research, a first priority would be to make a comparative study of the regeneration of such old regions as against the genesis of the new.

The second priority would be to study some of the consequences for employment and, thus, for socioeconomic character and lifestyle. In the United States, Bluestone and Harrison's work on the New England economy strongly suggests that high technology revival of an older industrial region simply leads to the deskilling of the old manufacturing workers and to an influx of new workers into the better new jobs (Bluestone and Harrison, 1980; 1983). Parallel work on Scotland or the Northwest, where the loss of traditional male manufacturing jobs has been particularly severe, would be instructive. But, again, it would be important to compare and contrast this with regions such as the M4 corridor and Cambridge, where the manufacturing base has been more modest—although similarly under great strain in recent years—and where the dominant employment traditions have been tertiary and white-collar in character. The British, having helped to pioneer research in this field, clearly have plenty yet to do.

REFERENCES

BAKER, K. (1982) "Information technology: industrial and employment opportunities." Royal Society of Arts Journal 130: 780-785.

BERRY, B.J.L. (1970) "The geography of the United States in the year 2000." Transactions of the Institute of British Geographers 51: 21-53.

BLACKABY, F. [ed.] (1979) De-Industrialization: National Institute of Economic and Social Research, Economic Policy Paper 2. London: Heinemann Education.

BLUESTONE, B. and B. HARRISON (1983) The Deindustrialization of America: Plant Closures, Community Abandonment, and the Dismantling of Basic Industry. New York: Basic Books.

———(1980) Capital and Communities: The Causes and Consequences of Private Disinvestment. Washington, DC: The Progressive Alliance.

BREHENY, M., P. CHESHIRE, and R. LANGRIDGE (1985) "The anatomy of job creation? Industrial change in Britain's M4 corridor," in P. Hall and A. Markusen (eds.) Silicon Landscapes. Boston: George Allen and Unwin.

BUSWELL, R.J. and E.W. LEWIS (1970) "The geographical distribution of industrial research activity in the United Kingdom." Regional Studies 4: 297-306.

CHECKLAND, S. (1975) The Upas Tree. Glasgow: Glasgow University Press.

DONNISON, D. and P. SOTO (1980) The Good City: A Study of Urban Development and Policy in Britain. London: Heinemann Education.

FOTHERGILL, S. and G. GUDGIN (1982) Unequal Growth: Urban and Regional Employment Change in the U.K. London: Heinemann Education.

FREEMAN, C. [ed.] (1984) Long Waves in the World Economy. London: Frances Pinter.

———C. CLARK, and L. SOETE (1982) Unemployment and Technological Innovation: A Study of Long Waves and Economic Development. London: Frances Pinter.

GERSHUNY, J. and I. MILES (1983) The New Service Economy: The Transformation of Employment in Industrial Societies. London: Frances Pinter.

GLASMEIER, A. K., P. HALL, and A. R. MARKUSEN (1983) Recent Evidence on High-Technology Industries' Spatial Tendencies: A Preliminary Investigation; Working Paper 417. Berkeley: University of California, Institute of Urban and Regional Development.

Great Britain Department of Trade and Industry (1983) Regional Industrial Development (Cmnd. 9111). London: HMSO.

Great Britain South East Economic Planning Council (1967) A Strategy for the South East: A First Report by the South East Planning Council. London: HMSO.

Great Britain South East Joint Planning Team (1970) Stretegic Plan for the South East: Report of the South East Joint Planning Team. London: HMSO.

HAKIM, C. (1981) "Job segregation: trends in the 1970s." Employment Gazette 89: 521-529.

——— (1979) Occupational Segregation: A Comprative Study of the Degree and Pattern of the Differentiation Between Men's and Women's Work in Britain, the United States, and Other Countries; Research Paper 9. London: Department of Education.

HALL, P. (1985) "The geography of the fifth Kondratieff," in P. Hall and A. Markusen (eds.) Silicon Landscapes. Boston: George Allen and Unwin.

HALL, P. and A. MARKUSEN [eds.] (1985) Silicon Landscapes. Boston: George Allen and Unwin.

HOWELLS, J. R. L. (1984) "The location of research and development: some observations and evidence from Britain." Regional Studies 18: 13-29.

JOSEPH, G. (1983) Women At Work: The British Experience. Oxford: Phillip Allan.

MALECKI, E. (1981a) "Public and private sector interrelationships, technological change, and regional development." Papers, Regional Science Association 47: 121-137.

——— (1981b) "Government-funded R&D: some regional economic implications." Professional Geographer 33: 72-82.

——— (1980) "Corporate organization of R&D and the location of technological activities." Regional Studies 14: 219-234.

MARSH, P. (1983) "Britain's high-technology entrepreneurs." New Scientist (November 10).

MASSEY, D. (1983) "Industrial restructuring as class restructuring: production decentralization and local uniqueness." Regional Studies 17: 73-90.

——— and R. MEEGAN (1982) The Anatomy of Job Loss. London: Methuen.

McQUAID, R. (1984) M4 Working Paper 1. (unpublished)

OAKEY, R. (1983) Reserch and Development Cycles, Investment Cycles and Regional Growth in British and American Small High Technology Firms: Discussion Paper 48. Newcastle upon Tyne: University, Centre for Urban and Regional Development Studies.

———— A. T. THWAITES, and P. A. NASH (1980) "The regional distribution of innovative manufcturing establishments in Britain." Regional Studies 14: 235-253.

Regional Studies Association (1983) Report of an Inquiry into Regional Problems in the United Kingdom. Norwich: Geo Books.

ROTHWELL, R. and W. ZEGVELD (1981) Industrial Innovation and Public Policy: Preparing for the 1980s and 1990s. London: Frances Pinter.

SAXENIAN, A. (1985) "The genesis of Silicon Valley," in P. Hall and A. Markusen (eds.) Silicon Landscapes. Boston: George Allen and Unwin.

TAYLOR, T. (1985) "High-technology industry and the development of science parks," in P. Hall and A. Markusen (eds.) Silicon Landscapes. Boston: George Allen and Unwin.

THWAITES, A. T. (1982) "Some evidence of regional variations in the introduction and diffusion of industrial processes within British manufacturing industry." Regional Studies 16: 371-381.

WORTHINGTON, J. (1982) "Industrial and science parks—accommodating knowledge based industries," in Planning for Enterprises: Proceedings of a Council of Europe Urban Policies Seminar. Swansea: City of Swansea.

Part II

The New Industrial Space

3

Innovative Manufacturing Industries: Spatial Incidence in the United States

AMY K. GLASMEIER

☐ THE CONTINUED DECLINE of basic industries in the manufacturing belt and the rise of new industries in other regions has led many observers to conclude that a shift is occurring in the location of industrial innovation (Rees, 1979; Norton and Rees, 1979). Factors thought to produce this shift include an uncompetitive industrial structure in the manufacturing belt (Chinitz, 1960), the cumulative impact of product cycle phases (Rees, 1979), government spending for military procurement and troop placement (Markusen, 1984; Weinstein and Firestine, 1979), and quality of life considerations (Hall, 1982; Rones, 1983).

This apparent shift has intensified competition between states for new high tech jobs and investment. Southern and Western states actively recruit firms from the Northeast and Midwest by offering them financial incentives and a better "business climate" (Harris, 1984). Northeast and Midwest states have responded with similar programs in an effort to remain competitive (Office of Technology

AUTHOR'S NOTE: *I am deeply grateful to Ann Markusen and Peter Hall for allowing me the opportunity to participate in the original National Science Foundation-sponsored project, Innovation and Regional Growth. I would also like to acknowledge Michael B. Teitz, who has been a guiding influence during my studies at the University of California at Berkeley; Richard Walker, Philip Shapira, Bennett Harrison, and Ann Lawrence for significant comments of drafts of this and other papers; Allen Sonafrank, Carlos Davidson, and the staff of the Qualitative Anthropology Laboratory for tremendous programming assistance; the staff of the Institute of Urban and Regional Development for their ongoing assistance; and to Barbara Baran and Tom Bell for their comments and continuing support.*

Assessment, 1983; Peltz and Weiss, 1984; National Governor's Conference, 1983). The race to offer a better business climate has resulted in the proliferation of state and local high tech programs, but to date no systematic evaluation of their effectiveness has been undertaken.

Few empirical studies of high technology industry location have focused on a state or regional level (exceptions include Armington et al., 1983; Brown, 1983; Glasmeier, Hall, and Markusen, 1983). This is due largely to the lack of suitable data sources to enable a detailed state-level industry analysis.

Using a well-known but infrequently analyzed data source, the 1972 and 1977 Census of Manufactures Plant location files and industry employment estimates, this chapter presents a state-level analysis of the location and growth of high technology industries in the mid-1970s.[1] The data base provides actual counts of manufacturing plants at a 4-digit industrial classification level for every county in the country. Employment estimates were constructed based on the plant-level data. The two have been merged and aggregated to metropolitan and state levels, allowing detailed analysis of high tech plant and employment location in the United States. (Glasmeier, Markusen, and Hall, 1983b). The chapter begins with a brief review of high tech definitions, and then presents preliminary and partial findings of state-level high technology industry employment.

DEFINING HIGH TECHNOLOGY INDUSTRIES

There is considerable debate about what constitutes a high technology industry. Although the computer and microelectronics industries are generally considered to be "high tech," the inclusion of other technology-intensive industries, such as chemicals and machinery, is controversial. A number of measures have been used to define high tech, all of which lack precision and comparability (Vinson and Harrington, 1979; Technical Marketing Associates, 1979). Specifically, the measures either are too aggregate or fail to treat all industries in the same manner. For the purpose of this chapter, a working definition is based on the human capital component of the labor process. That is, industries are considered high tech on the basis of the proportion of technical occupations in the total industry labor force (Glasmeier, Markusen, and Hall, 1983a).

Any definition of high tech industry must take account of the constantly changing nature of technology. This suggests that there can be no temporally static list of high tech industries. Given that new technologies are always being incorporated into most industries, the technical component of the overall labor force is rising (Baran, this issue). Using a definition based on occupations controls for this effect, but necessitates a revision of the criterion used to select particular industries.

Most lists of high tech industries are based on the Standard Industrial Classification (SIC) system and include parts of industry groups such as chemicals, electrical machinery, transportation equipment, communications equipment, and engineering and scientific instruments.[2] As part of my initial analysis, two definitions of high tech were applied. The more restrictive definition uses the national durable manufacturing industry average for scientists, engineers, and technicians (Green et al., 1983); the broader uses the overall manufacturing average for engineers, engineering technicians, computer scientists, mathematicians, and life scientists, including geologists, physicists, and chemists (Glasmeier, Markusen, and Hall, 1983a). The computations using both definitions were similar and, therefore, only the results of the broader definition are presented here. Table 3.1 lists the industries identified using the two cut-off points, showing the similarities and differences between the two groups.

THE LOCATION OF HIGH TECHNOLOGY INDUSTRIES AND EMPLOYMENT

The intense interest in high technology industries overstates their actual contribution to total employment. In 1977 high tech employment accounted for approximately 5% of all employment and 24% of manufacturing jobs. In comparison to growth in non-manufacturing sectors, high tech industries have added relatively few jobs (Brown, 1983), and high tech employment, although increasing rapidly relative to other sectors, is expected to contribute only 10% of all new jobs created over the next 10 years (Richie et al., 1983).

In addition, because most of the evidence about high technology industry location has been anecdotal, it is now common wisdom that the South and the West have a disproportionate share of high tech industries and, therefore, high tech employment. On the basis of two measures at least, the location of high tech plants and the absolute concentration of employment, the emperical evidence contradicts this popular imagery.

TABLE 3.1
Employment in High Technology Industries

Definition I

SIC Code	Industry Name	Estimated Employment
2812	Alkalies and chlorine	11832
2813	Industrial gases	7397
2816	Inorganic pigments	12000
2819	Ind. inorganic chem. nec.	78203
2821	Plastic materials, syn. resins	57106
2822	Synthetic rubber	11545
2823	Cellulosic manmade fibres	16229
2824	Synthetic org. fibres, ex cellul	74064
2831	Biological products	18468
2833	Med. chem. botanical products	15725
2834	Pharmaceutical preparations	126399
2861	Gum, wood chem.	4720
2865	Cyclic crudes, intermediates, dyes	35513
2869	Ind. organic chem. nec.	112400
2891	Adhesives, sealants	16671
2892	Explosives	11549
2893	Printing ink	10100
2895	Carbon black	2600
3899	Chem. chem. prep., nec.	35298
2911	Petroleum refining	102399
3482	Small arms ammunition	12187
3483	Ammunition, ex small arms, nec.	20581
3484	Small arms	17499
3489	Ordnance, accessories, nec.	19037
3511	Steam, gas, hydraulic turbines	40964
3519	Internal combustion engines, nec.	88799
3531	Construction mach. equipment	155198
3532	Mining mach., equipment	31298
3533	Oil field mach., equipment	58499
3534	Elevators, moving stairways	10201
3535	Conveyors, conveying equipment	32926
3536	Hoists, ind. cranes, monorail system	15799
3537	Ind. trucks, tractors, trailers, stackers	28385
3561	Pumps, pumping equipment	63055
3562	Ball, roller bearings	50288
3563	Air, gas compressors	31900
3664	Blowers, exhaust, ventil. fans	28430
3565	Ind. patterns	9398
3566	Speed chgers, ind. high drives, gears	24546
3567	Ind. process furnaces, ovens	16263
3568	Mech. power transmission equipment nec.	32558
3569	General ind. mach. equipment nec.	58554
3573	Electronic computer equipment	192513
3574	Cal. acc. mach. ex elec. computer equipment	15459
3576	Scales, balances, ex. lab.	6712
3579	Office mach., nec.	42412
3621	Motors, generators	96970
3622	Industrial controls	56428
3623	Welding apparatus, electric	17399
3624	Carbon, graphite products	12082
3629	Elec. ind. apparatus, nec.	16474
3661	Telephone, telegraph apparatus	124309
3662	Radio TV transmit, signal, detect. equipment	332923

TABLE 3.1   Continued

SIC Code	Industry Name	Estimated Employment
3671	Cathode ray tubes, nec.	36799
3674	Semiconductors, related devices	114000
3675	Electronic capacitors	28643
3676	Resistors for electronic app.	24922
3677	Resistors, electric apparatus	22425
3678	Connectors, electronic appls.	26013
3679	Electronic components, nec.	125965
3721	Aircraft	222799
3724	Aircraft engines, parts	106200
3728	Aircraft Parts, auxiliary equipment nec.	101934
3761	Guided missile, space vehicles	93928
3764	Guided missile, space vehicle propulsion units	17013
3769	Guided missile, space vehicle parts nec.	10193
3811	Eng. lab. scientific research inst.	42197
3822	Auto controls reg. resid. comm. env. appl	39099
3823	Ind. instr. measure, display	46499
3824	Fluid meters, counting devices	16018
3825	Instr. measuring, testing, elec. sigs.	66600
3829	Measuring, controlling devices, nec.	32200
3832	Optical instruments, lenses	29905
3861	Photographic equipment, supplies	111556
Definition I:	Total Employment	3835211

A Broader Definition Includes:
Definition II

SIC Code	Industry Name	Estimated Employment
2851	Paints, varnishes, lacquers, enamels	61297
2844	Perfumes, cosmetics, toilet prep.	50799
2843	Surface active finishing agents	6838
2842	Spec. cleaning, polishing prep.	22919
2841	Soap, other detergents	32641
3561	Pumps, pumping equipment	63055
3549	Metalworking machinery, nec.	19141
3547	Rolling mill mach., equipment	8530
3546	Power driven hand tools	27676
3545	Mach. tools access., measuring devices	54257
3544	Spec. dyes, die sets, jigs fix., ind. molds	106107
3542	Mach. tools, metal forming types	23145
3541	Mach. tools, metal cutting types	59463
3652	Phono records, prerecorded mag tapes	23102
3651	Radio, TV receiv. sets, ex comm. types	74600
3743	Railroad equipment	56399
3843	Dental equipment, supplies	16636
3842	Orthopedic, prosthetic, surgical appliances	53990
3841	Surgical, medical instr. apparatus	43225
Definition II:	Estimated Employment Total	4760506

SOURCE: Both definitions are based on the Occupational Employment Statistics Survey, 1980, Department of Labor. *Definition I:* Vinson et al. (1983); *Definition II:* Glasmeier et al. (1983a).

TABLE 3.2
The 13 Largest High Technology Sectors, 1977
(all numbers are absolute)

Industry Name (SIC Code)	1977 Estimated Employees	# of States with the Industry 1972	# of States with the Industry 1977
Radio TV transmit, signal, detect. equipment (3662)	332,923	48	50
Aircraft (3721)	222,799	33	34
Construction mach. equipment (3531)	155,198	46	47
Electronic comput. equipment (3573)	192,513	40	45
Pharmaceutical preparations (2834)	126,399	41	43
Electronic components, NEC (3679)	125,965	44	48
Telephone, telegraph apparatus (3661)	124,309	37	40
Semiconductors, related devices (3674)	114,000	37	42
Ind. organic chem. NEC. (2869)	112,400	41	41
Photographic equipment, supplies (3861)	111,556	42	41
Aircraft engines, parts (3724)	106,200	28	33
Petroleum refining (2911)	102,399	41	39
Aircraft parts, auxiliary equipment, NEC. (3728)	101,934	40	43
Totals	1,928,595		

SOURCE: Glasmeier (1985).
NOTE: NEC = not elsewhere classfied.

In fact, high tech *plants* are found in 80% of the states; for example, plants in the communications equipment industry (SIC 3662), the largest single industry in terms of employment, are found in all 50 states. No doubt this reflects the general diversity found in this particular 4-digit product group. But the widespread distribution of these industries further suggests that many are market-oriented and produce highly differentiated products with considerable linkages to other high tech sectors. Another explanation for this dispersion is the spatial consequences of different phases of individual industry product cycles. Thus, for example, while new product development of semiconductors is concentrated in states such as California,

TABLE 3.3

The 13 Largest 4-Digit High Technology Industries, 1977

% of Total Nat'l Industry Employ. Acc'ted for by Indiv. States	Standard Industrial Classification Code of 13 Industries												
	3662 83%	3721 90%	3531 89%	3573 85.6%	2834 84%	3679* 82.9%	3661 82%	3674 97.5%	2869 85%	3861 92%	3724 75%	2911 72.5%	3728 86.3%
AR	2.0						3.4						
AZ				4.3				8.2			4.9		36.0
CA	23.0	27.0	4.0	29.0	7.0	23.0	6.0	31.0	3.0	4.0	3.0	15.0	31.0
CN	2.2	6.0			1.0	4.0	1.0	4.0	1.0				1.0
CO	2.0			1.6			3.0			10.0			
FL	7.6	6.1		4.0		4.0		4.0		4.0	1.0	1.0	5.7
GA		5.9	1.0										
IA	1.0		14.0										
IL	3.9		21.0		8.2	10.9	20.6		1.7				7.0
IN	2.0		3.0		7.0	1.0	5.0		1.0	1.0	10.0	3.0	4.0
KS	1.0	12.0	1.0										
LA						2.9			15.1				9.5
MA	6.0			12.4	1.0	7.0	4.0	4.0	1.0	15.0	10.0		
MI	1.0		6.0		5.0	1.0			3.7	1.0	1.0	2.2	
MN	2.0		4.0	9.5		6.5					9.7		
MO	1.1		1.0		2.1	3.2				2.4			
MS							2.9						
NC		3.1			6.2		4.0			4.9			

(continued)

61

TABLE 3.3 Continued

% of Total Nat'l Industry Employ. Acc'ted for by Indiv. States	3662 83%	3721 90%	3531 89%	3573 85.6%	2834 84%	3679* 82.9%	3661 82%	3674 97.5%	2869 85%	3861 92%	3724 75%	2911 72.5%	3728 86.3%
NJ	6.6			4.0	18.0	6.0	3.0	4.0	7.0	8.0	1.0	5.0	2.0
NM							2.9						
NY	11.7	6.0	2.0	8.3	1.5		2.9	12.5	4.0	30.0	6.0		
OH	2.0		11.0		2.7	2.3		4.0	2.6		11.8	3.0	3.0
OK			3.0	2.5			3.4					3.6	9.1
PA	2.8	5.8	5.0	4.0	12.0		7.0	10.8	4.0	1.0	6.0	8.0	7.0
TX	5.7	12.5	4.0	5.0	1.0	2.0	4.0	16.0	26.0			27.0	2.0
UT				1.5						6.7			
VA							6.1		3.6				
VT								3.0			4.4		
WA		5.8		1.0									4.8
WI	1.0		11.0										
WV									9.3				

SOURCE: Glasmeier (1985).

NOTE: SIC numbers for industries as follows: 3662 = Radio, TV, transmit, signal, detec, equip.; 3721 = Aircraft; 3531 = Construction machinery, equip.; 3573 = Electronic computer equip.; 2834 = Pharmaceutical preparations; 3679 = Electronic components, NEC; 3661 = Telephone, telegraph apparatus; 3674 = Semiconductors, related devices; 2869 = Ind. organic chem., NEC; 3861 = Photographic equip., supplies; 3724 = Aircraft engines, parts; 2911 = Petroleum refining; 3728 = Aircraft parts, auxiliary equip., NEC; NEC = Not elsewhere classified.

TABLE 3.4
Estimated Employment in High Technology Industry, 1977

States	
California	641,262
Illinois	360,317
New York	336,848
Pennsylvania	314,322
Ohio	295,135
Texas	285,667
New Jersey	232,268
Massachusetts	204,630
Michigan	169,400
Connecticut	160,085
Indiana	153,458
Wisconsin	128,809
Florida	119,048
Percent total	71%
Total	3,401,259

SOURCE: Glasmeier (1985).

Arizona, and New York, assembly operations occur in these and other states and also in other countries (Saxenian, 1981).

On the other hand, although high tech industries are found in many states, employment is highly concentrated. First, as many as 100 industries make up the high tech group; nevertheless, the great majority of all high tech manufacturing employment is concentrated in only a few industries. Just 13 contribute approximately 1.9 million jobs to the economy and represent about 50% of all high tech employment (Table 3.2)

Second, fewer than 17 states are responsible for a majority of high tech jobs in the largest industries. The regional distribution of this employment includes states in the Middle Atlantic, the Great Lakes, the Northeast, the Southwest, and the Pacific regions. Despite widespread growth of high tech employment since 1977 (the year corresponding to the data base analyzed here), recent analysis of state-level employment confirms that the majority of jobs are still concentrated in a few states (see U.S. Department of Commerce, 1983, for details; also Table 3.3)

States in the Northeast, Middle Atlantic, and Great Lakes regions account for 10 of the top 13 states in terms of total high tech plants and employment (see table 3.4). These 13 are collectively responsible for 71% of total high tech employment.

The prominence of Northeast and Midwest states within this group, although notable, is not surprising given their share of overall manufacturing employment. The concentration of high tech

TABLE 3.5
Percentage of Manufacturing Employment
Accounted for by High Tech Jobs in 48 States and
the District of Columbia

State	Total Manufacturing	State	Total Manufacturing
Arizona	41	Indiana	21
Connecticut	39	Arkansas	20
California	37	Missouri	20
Kansas	37	Virginia	26
Colorado	35	South Carolina	19
Massachusetts	33	South Dakota	18
Oklahoma	33	New Hampshire	18
Texas	32	Delaware	17
Utah	31	Kentucky	17
Florida	31	Tennessee	17
New Jersey	30	Washington	17
Maryland	30	Rhode Island	16
Louisiana	29	Michigan	15
Illinois	28	Alabama	14
Minnesota	27	Idaho	14
West Virginia	27	Oregon	12
New Mexico	26	North Carolina	12
Wisconsin	24	Mississippi	11
Pennsylvania	23	Georgia	10
Nevada	23	Maine	9
Iowa	23	Montana	9
New York	22	Alaska	.05
Ohio	22	Hawaii	.03
Wyoming	22	Washington D.C.	.01
Nebraska	21	National	.24

SOURCE: Glasmeier (1985).

employment in Northeastern and Midwestern states also correlates with research findings on the location of corporate research and development activities. Even when controlling for size, Malecki (1979) found that a significant correlation exists between manufacturing employment and the concentration of research and development employment and activities.

PROPORTIONAL MEASURES OF HIGH TECH EMPLOYMENT

Based on the absolute number of jobs then, Northeast and Midwest states have benefitted from the growth of new technology industries. In fact, in terms of absolute employment, many of the high tech "winners" are located in the "frostbelt." But sheer numbers of jobs tend to mask the structural differences found in a state's manufacturing base; it is important, therefore, to examine the proportion of high tech jobs to total manufacturing employment. Such

TABLE 3.6
High Tech Centers:
State Location Quotients Greater Than Unity, 1977

State	Location Quotient
Arizona	1.80
Connecticut	1.65
Kansas	1.63
Colorado	1.57
California	1.49
Massachusetts	1.43
Florida	1.39
Oklahoma	1.38
Texas	1.33
Utah	1.33
New Jersey	1.23
Maryland	1.22
Louisiana	1.22
Minnesota	1.18
Illinois	1.15

SOURCE: Glasmeier (1985).

a measure highlights the extent to which a state's manufacturing base
in composed of innovative industries (Table 3.5)

Comparing state high tech levels to total manufacturing employment indicates that only 18 states have a high tech base exceeding the
national average (Table 3.5). According to this measure, the ranking
among states varied considerably from that based on absolute
numbers of jobs. Only 4 states (Texas, Connecticut, California, and
Massachusetts) had similar rankings on both absolute number of jobs
and on the ratio of high tech jobs to total manufacturing employment.
Few Northeast or Midwest states (6) ranked high on this measure,
suggesting their industrial base may be less technically innovative
than others.

Because simple proportional measures of employment concentration do not fully account for size differences in overall state
economies, location quotients (LQ) were calculated in order to
identify states that have a disproportionate share of high tech
employment to total manufacturing employment. That is,

$$LQ = [HTEMP_i / \Sigma HTEMP] / [SMFT_i / \Sigma SMFT]$$

where $HTEMP_i$ is total high tech employment for 1977, in state i (i =
1...n) and SMFT is state manufacturing employment (Table 3.6).

A location quotient provides a simple method for comparing a
state's percentage share of a particular economic activity with its
percentage share of some base aggregate (Isard, 1960). In this case we

compare a state's share of high tech manufacturing employment to its share of national manufacturing employment. This measure provides a rough assessment of a state's strength in industries using the latest technological innovations, and it allows us to highlight states with concentrations of high technology manufacturing employment.

A location quotient of 1 indicates that the ratio of a state's share of high tech employment to total manufacturing is equivalent to the distribution of high tech and manufacturing employment nationally. A value greater than 1 indicates a state that specializes in these industries. Using a location quotient allows us to identify states that might be considered centers of technical innovation.

According to this measure, fewer than 30% of the states have a technically innovative manufacturing base. In addition, the regional distribution of states with high LQs does differ from that based on the absolute levels of high tech jobs. States in the South and West are represented more frequently in this group. Nevertheless, the difference is small. Of the 13 largest states with over 500,000 manufacturing jobs, half had LQs greater than 1. Of these, 4 were in the Northeast and Middle Atlantic states and 1 was in the Great Lakes region. If the Plains region is also included, the difference between the frostbelt and the sunbelt disappears. Thus, we cannot conclude on the basis of this measure that either high tech employment or innovative technical activity is concentrated outside the manufacturing belt in states with a relatively new history of manufacturing development.

HIGH TECH EMPLOYMENT CHANGE

At the same time, however, there does seem to be a difference between the sunbelt and frostbelt in terms of new high tech employment growth. Changes in employment growth during the mid-1970s suggest that the location of new manufacturing jobs may reduce the concentration in the Northeast and Midwest (Armington et al., 1983). New high tech growth, as opposed to in-place employment, has been the focus of attention in the economic development literature (Florida Department of Commerce, 1983; Glasmeier et al., 1984). States not otherwise known as manufacturing centers have demonstrated a capacity to attract new high tech jobs. Depending on the magnitude of this change, the differential growth in high tech jobs may indicate that new centers of technical innovation are developing.

In the mid-1970s a great majority of states gained both new high tech industries and additional high tech jobs. On average, state-level high tech employment increased by 17%. Of the 20 states that experienced an increase greater than the average, 14 were in the South and West. This regional concentration is somewhat misleading, however, given that these states had a limited high tech base to begin with so that small absolute changes result in large percentage gains. Only Texas, North Carolina, and Minnesota were found both to have a substantial high tech employment base and to experience a considerable percentage increase in new high tech jobs.

During the 1970s manufacturing industries increased by a mere 3.7%. This in part reflects a post-Vietnam war decline in demand for military-related products, as well as the impact of the major recession in 1974-1975. High tech employment, although only a modest portion of total manufacturing employment, was important in the overall rate of growth. In only a few states (Kansas, Wisconsin, Wyoming, and South Dakota) was the percentage change in manufacturing employment greater than percentage change in high tech manufacturing.

States such as New York, Illinois, West Virginia, and New Jersey actually gained high tech jobs despite an overall decline in manufacturing employment. Other states, such as Ohio, Pennsylvania, Delaware, Maryland, and Washington, D.C., experienced the opposite trend of percentage declines in both high tech employment and total manufacturing. Nonetheless, in absolute terms, the top 13 states gained 47% of net new high tech employment created in the mid-1970s. Although such growth is impressive, the top 13 did not gain a comparable share of national high tech job growth based on their absolute share of overall high tech employment. In the same sense, then, although the evidence is limited, there is some indication that in the future the South and West may, indeed, gain a disproportionate share of new high tech jobs.

HIGH TECH INDUSTRY DIVERSITY AND SECTOR EMPLOYMENT CONCENTRATION WITHIN STATES

HIGH TECH INDUSTRY DIVERSITY

The dramatic decline of basic industrial employment raises serious questions about the vulnerability of regional economies that are dependent on a few mature industries. This vulnerability is exacerbated by the large size of individual production units characteristic of many mature industries; for example, the closure of

TABLE 3.7
High Tech Industry Diversity:
Number of Industries in Each State, 1977

State	Number of Industries	State	Number of Industries
California	100	Oklahoma	76
Ohio	98	Maryland	74
Texas	98	Kentucky	71
Pennsylvania	97	Louisiana	71
New York	96	Arizona	70
Illinois	93	Rhode Island	69
Florida	92	Utah	67
Michigan	92	New Hampshire	66
Indiana	91	Mississippi	63
New Jersey	91	Arizona	62
Massachusetts	90	Nebraska	62
Wisconsin	90	West Virginia	62
Tennessee	89	New Mexico	52
Connecticut	88	Maine	48
Missouri	86	Nevada	44
Alabama	85	Delaware	40
Colorado	85	Vermont	38
Minnesota	85	Idaho	35
North Carolina	83	South Dakota	32
Washington	83	North Dakota	29
Georgia	82	Montana	27
Iowa	82	Hawaii	24
Virginia	82	Wyoming	21
Kansas	79	Washington, D.C.	16
Oregon	79	Arkansas	11
South Carolina	77		

SOURCE: Glasmeier (1985).

one steel plant translates into the loss of several thousands of jobs. A closure of this magnitude ripples through an economy as the demand for goods and services supported by the income paid to employees declines. In light of this experience, states and communities have initiated industrial development programs to diversify their industrial bases. The rapid growth of many high tech industries over the last several years makes them important candidates for such diversification efforts.

Contemporary high tech centers conjure up images of industry clusters linked together as both inputs and outputs to form new industrial complexes. States with well-known centers, such as Silicon Valley in California and Route 128 in Massachusetts, are important but rare. Indeed, few states are thought to have a diversified high tech base (Fantas Corporation, 1983).

The actual distribution of high tech industries across states is quite broad. Several states (39) have more than 50 of the 100 industries studied. States with relatively few high tech industries were located

TABLE 3.8
Industries with Estimated Average Plant Size
Greater Than 300 Employees, 1977

Industry		Average Number of Employees/Plant
3761	Guided missile, space vehicle	2348
3721	Aircraft	1265
2824	Synthetic organic fibers, ex cellul	1122
3764	Guided missile, space vehicle propulsion units	654
2823	Cellulosic manmade fibers	649
3795	Tanks, tank components	505
3511	Steam, gas, hydraulic turbines	493
3661	Telephone, telegraph apparatus	470
3724	Aircraft engines, parts	394
3519	Internal combustion engines, nec.	383

SOURCE: Glasmeier (1985).

outside the traditional manufacturing belt. These undiversified states topped the list for percentage gains in high tech employment and were found to be mainly in the Southeast, Southwest, and Mountain states (Table 3.7).

The composition of high tech employment varies across states and appears to be correlated with size of manufacturing base and employment level. For analytical purposes, in the remaining sections states are grouped into three categories—small, medium, and large—based on their total manufacturing base.

The 13 largest states (those with more than 500,000 manufacturing jobs) had a diverse high tech base. At least 90% of the industries were found within their borders. The distribution ranged from all 100 industries in California to 88 in Connecticut. For the largest states, having a diversified high tech base is primarily related to overall size. Large states with a diversified industrial structure and existing industrial agglomerations are likely candidates for continued industrial diversification, including high tech industries. In a limited sense, the most populated states still serve as seedbeds for innovative activities, even though their share of any one industry's employment may be declining.[3]

Medium-sized states (with between 100,000 and 500,000 manufacturing jobs) exhibit two distinct trends. The first group has a high proportion of high tech employment to total manufacturing and has a correspondingly high ranking on number of high tech industries present. The second group has relatively low levels of high tech employment to total manufacturing and, in general, has fewer high tech industries. Coincidentally, the latter also tends to be composed of the largest states within the medium-sized group.

The regional composition of medium-sized states also varied considerably. The first group was composed of Southeast, Southwest, and Western states. Almost half the states in the other group were also located in the South. Southern states' manufacturing industries appear to share the characteristic of being both high and low tech.

INTRASTATE SECTORAL CONCENTRATION

The widespread presence of high tech industries in both large and medium-sized states tends to support the proposition that the growth of high tech industries may reduce a region's vulnerability to unexpected declines in any one sector. The in-state concentration of high tech employment in a few key industries, however, highlights an important countervailing trend. In 31 states 5 industries accounted for more than 50% of total high tech employment. This sectoral concentration was most prevalent in the medium and smaller states, with the exception of Texas and Wisconsin.

The largest manufacturing states have a higher degree of industrial diversity and a correspondingly lower level of sectoral concentration. In medium-sized states, again, two divergent trends occur. States with high percentages of high tech employment to total manufacturing had high levels of sector concentration. Only Connecticut had less than 50 percent of its high tech employment concentrated in 5 industries. The other group of medium-sized states showed much less sectoral concentration; only 4 of the 11 states had concentrations exceeding 50%. Small states show less diversity and have concentrated employment.

The unexpected finding that high tech industries are widely dispersed but in-state employment is concentrated points to the possibility that states may be trading one form of economic dependence for another. A specialization in particular industries does not necessarily mean an area is dependent, however, unless perhaps the employment is concentrated in large production facilities.

AVERAGE PLANT SIZE

The favorite image of a high tech establishment is that it is run by an ingenious engineer who leaves a large company to strike out on his or her own (Deutermann, 1966; Koch et al., 1983). This image, although attractive, does not fit with the data on actual size distribution of high technology establishments. High tech plants are considerably larger (92 employees) than the national average (54).[4]

For example, 25 of the 100 industries have an average plant size exceeding 200 employees; 40 averaged over 100. Three industries in particular (aircraft, missiles, and synthetic fibers), all have an average plant size exceeding 1000 employees. These 3 industries benefit from economies of scale; aerospace industries on the basis of low-volume, large-scale, single-unit products, while synthetic fibers is a high-volume, continuous-processing industry.

The distribution of high tech average plant size shows considerable variation across states. Of the largest states, 62% had an average plant size exceeding that of the nation. Examples include aircraft production in Pennsylvania, New York, California, and Texas; communications equipment and drugs in New Jersey, Pennsylvania, and New York; aircraft engines in Ohio and Massachusetts; and radio and television equipment in Illinois.

Medium-sized states again show two distinct patterns. Of the states with a large high tech employment base, 75% had large plants. Four states had large aircraft and missile plants (Georgia, Connecticut, Florida, and Washington), three had large chemical complexes (Louisiana, West Virginia, and Virginia), and Minnesota had several large computer production facilities.

The second subgroup of medium-sized states, those with a less-developed high tech base, had smaller plants. Fewer than half of these states' average plant size exceeded the national average. In 3 of the remaining 5, aerospace production facilities explained the large plants.

Without exception, small states had smaller than average-sized plants. This can be attributed to the low level of high tech industry found within them. It is also possible that the structure of the firms in these states differs from others. This may be explained by the fact that in high tech industries the importance of a technical labor force and the need for proximity to headquarters mitigates against large firms locating in remote areas. Entrepreneurs, on the other hand, have a great propensity to locate in or near their hometown. Consequently, high tech industries found in the smaller states are more likely to be locally-owned ventures than branch units of a large corporation (Harris, 1984). On the other hand, smaller plants also may indicate higher rates of capital intensity associated with new production facilities.

State and local policies designed to attract high tech industries are premised on the belief that these industries will increase local economic diversity and, therefore, reduce future vulnerability to decline in more mature industries. The widespread distribution of high tech industries suggests that they are good candidates for strategies aimed at promoting economic diversification. But the presence of a number of industries clustered around a few large employers signifies that an area may be more vulnerable to changes in one sector's employment. Their vulnerability depends on factors influencing a product's life cycle, such as domestic and foreign competition, the maturation of individual products, and the introduction of automated processes in production facilities. For the most part, these factors are beyond the influence of an individual state or community.

HIGH TECH CENTERS

The success of a few states in establishing high tech industry bases has led many others to believe that they too can have a "slice of the high tech pie." States with a mature industrial base see their future tied to the development of new high tech industry clusters. The transformation of the Massachusetts economy from old to new industries is held up as an example of how this transition can be made. On the other hand, states in the sunbelt base their future on being able to attract production facilities that will eventually create concentrations of self-reinforcing innovative industries. Such optimistic assessments ignore the fact that very few states actually enjoy substantial concentrations of high tech employment.

Fourteen states were identified as centers of technical innovation based on their location quotient scores. One factor left unexplored is the type of production found in these states. The designation "innovation center" ignores whether production in these states is routine. Conceivably, any one of these states could be simply a production location, leaving the reader to conclude that it is really not an innovative center in the true sense of the word.

Using the 1977 Occupational Employment Statistics matrix for states and the nation, six 2-digit industries were examined.[5] Comparisons were made of the percentage of engineers in individual industries for both states and the nation. Data were not available for all states or for all 2-digit industries in each state. In each state comparison only industries for which 2-digit data were available were examined. Conclusions are made on the basis of a subset of both states and industries. Particular emphasis is placed on the 14 states considered high tech centers based on location quotient scores. Of these, 13 reported occupational data for one or more industries. Other states were also examined and the results are reported here. The reader should keep in mind that the results are partial due to data limitations.

Although data were available for 32 states, only 15 had one or more 2-digit industries in which the percentage of engineers exceeded the national average. Of the 15, 11 were states with location quotients greater than 1. Indeed, only 2 high tech centers failed to exceed the average for one or more industry, Oklahoma and Illinois. Oklahoma's poor performance may be due to measurement error, while Illinois no doubt faltered because of its diverse industrial base, which includes both high and low tech industries. More disaggregated data are needed to confirm this.

Five other states had a high proportion of engineers in one or more industries studied, but were not considered centers of innovation because of their low scores on the location quotient. This group included New York, New Mexico, Oregon, Utah, and Indiana. Of the 5, all but New York had an LQ equal to 1.0, indicating that their relative shares of national manufacturing employment and high tech employment were of equal proportions.

It is difficult to generalize these findings given the lack of data for all states and all industries within states. Still, the results indicate that states identified as centers of technical innovation have a substantial technical labor force found in conjection with existing high tech agglomerations. The high levels of engineers in several industries in New York no doubt reflects the existence of corporate research and development activities. The 4 states with location quotients approaching unity may be up-and-coming centers of innovation. Analysis of more recent data might clarify this trend.

A strikingly consistent characteristic of these so-called innovation centers is their dependence on defense industries, a dependence dating back to the Korean war (Bolton, 1966).[6] In research on the relationship between high tech industries and military expenditures, Markusen (1984) reports of significant market relations between high tech industries and defense department procurement. Making this point vivid, 10 of the 14 innovation centers had at least one, and 5 had two, defense sectors, making up a large portion of their high tech employment base.

Some states that have historically depended on defense spending, such as California, Texas, and Arizona, have successfully made the transition from being defense centers to being areas dominated by producers of new commercial technologies. (Some would argue that their historical dependence has not changed much.) Others, however, such as Connecticut, Kansas, Oklahoma, Colorado, and Florida, remain heavily dependent on defense contracting. For example, although Florida has well over 100,000 high tech jobs, two employers—the Harris Corporation and Martin Marietta—account for over 25% of the state's high tech employment.[7] An interview with corporate officials in the Harris Corporation revealed that defense-related government contracting accounted for between 75% and 85% of their business. Furthermore, the Defense Department is underwriting much of the corporation's research and development spending through a cooperative program.

LOCATIONAL TENDENCIES OF HIGH TECH
INDUSTRIES
AT A METROPOLITAN LEVEL

The state-level analysis of the location of high tech industries masks important differences that may exist within individual states. A companion study by Glasmeier, Hall, and Markusen (1983) on the spatial tendencies of high tech industries at a metropolitan level confirms many of my state-level observations.[8] Specifically, few metropolitan areas within states have substantial concentrations of high tech employment. Furthermore, high tech employment in a large number of metropolitan areas is concentrated in a few industries. These results reinforce the findings that there are very few centers of innovation and their high tech base is dominated by a handful of high tech industries.

But a number of interesting differences between the state and metropolitan levels highlight the complex nature of high tech employment growth. Despite an overall increase in high tech employment during the mid-1970s, many metropolitan areas did not share in this growth. Almost one-third of the 264 metropolitan areas studied actually lost high tech jobs. Although the losers were found in all regions, Northeast and Midwest metropolitan areas appeared to suffer the greatest losses. For example, two metropolitan areas in New York state (Syracuse and New York City) together lost an estimated 14,000 jobs.

On the other hand, metropolitan areas experiencing the greatest employment increases reflect those that are popularly considered high tech centers, such as Boston, San Jose, Dallas, and Phoenix. Nevertheless, these states also had their share of metropolitan high tech employment losers. For example, one-quarter of the metropolitan areas in California, an eighth of those in Texas, and half of those in Massachusetts lost high tech employment during the mid-1970s. Only Arizona, with only two metropolitan areas, showed consistent gains over the period. It is important to remember then that even within so called "winning states" the growth of high tech employment is unevenly distributed across metropolitan areas.

From the preceding state-level analysis we saw that states appear to specialize in selected high tech industries. At such an aggregate level, however, this specialization is open to a variety of interpretations. At a metropolitan level, however, industrial specialization has important implications for long-run economic stability. As Glasmeier et al. (1984) point out, in more than half of the metropolitan areas studied, high tech employment was concentrated in 5 or fewer industries. In fact, in 75 metropolitan areas, just one industry accounted for an estimated 70% of total high tech employment.

This evidence suggests that a number of the relationships found to occur across states are replicated at a metropolitan level. Many metropolitan areas host one or a few large employers and a small number of other high tech industries. In Melbourne, Florida, for example, one large employer has dominated the local economy since the early 1960s. Although the firm employs almost 4,000 engineers, the classified or defense-related nature of its products has actually inhibited the development of industrial spinoffs. This suggests that the growth potential of high tech industrial complexes is significantly influenced by individual products and markets. Thus, gaining a share of high tech employment does not guarantee that a Silicon Valley or a Route 128 will eventually be created.

CONCLUSIONS

The distribution of high tech industries and employment provides a rather surprising picture of spatial diversity: Plants are widely dispersed but employment is concentrated; industries are widely represented, but the majority of employment is really concentrated in only a few sectors; and sectoral concentration at a state level can be explained primarily by large production establishments.

The regional distribution of high tech employment also shows great diversity. Northeast, Middle Atlantic, and Great Lake states are well represented in terms of absolute number of jobs. Although these states perform less well on the basis of other measures, several have a highly innovative manufacturing base. Significantly, only a few states within each region have been substantial beneficiaries of high tech employment growth. While a marginal change in the distribution of high tech employment is occurring in other states, this shift is relatively small in comprison to the total, and the type of jobs being created appear to be nontechnical and perhaps routine in nature.

The concentration of high tech employment in the 14 states identified as technical innovation centers did not develop overnight. Researchers in the late 1950s (Perloff et al., 1960) saw the foundations forming for the development of new industrial complexes. For example, the authors proposed that the rapid growth of the aerospace industry in California (and missile production in particular) and its heavy use of electrical machinery would result in the development of a sophisticated electronics industry in Los Angeles. By this and other accounts, government programs, such as NASA and military weapons systems, are key in explaining the historical origins of many of these complexes.

The competiton between the frostbelt and the sunbelt ignores the strength of the manufacturing belt in absolute numbers of high tech jobs. Shifts that are occurring may eventually help to equalize the distribution of manufacturing jobs nationally. New England, California, and Texas still maintain a considerable lead in the distribution of technical employment within these industries. Only a redistribution of research and development functions will alter this pattern.

The link between technical innovation and high tech industries is poorly understood. Gaining high tech jobs does not automatically translate into the development of a technical manufacturing base. Key ingredients such as research and development activities, administrative functions attached to corporate headquarters, and advanced producer services are still concentrated in the Northeastern and Midwestern states (Stanback and Noyelle, 1984). The link between high tech industries and defense expenditures is also poorly understood. Several sectors, such as computers, semiconductors, and scientific instruments, rapidly evolved into commercial industries. Others, such as aircraft and ordinance remain almost totally military-oriented. This dichotomy translates into different development paths for states and communities. Further examination of interindustry linkages among different high tech sectors is needed in order to understand the long-term consequences of regional growth based on high tech industries.

NOTES

1. More recent data would add considerably to this analysis but, unfortunately, the results of the 1982 Census of Manufactures were unavailable at the time of printing.

2. The SIC code is a numerical scheme that classifies into groups industries and products; the larger the "digit," the more industry detail presented.

3. The possible seedbed effect is partially verified by Harris (1984), who shows that large Northeastern metropolitan areas have a higher ratio of independent high tech firms compared with branch affiliates of local or nonlocal corporations.

4. Armington et al. (1983) found that high tech establishments consisted of nonlocal affiliates and were larger than the national average of all establishments.

5. The 2-digit industries include SIC 28, 34, 35, 36, 37, 38.

6. Although beneficial in the current period, defense dependence can have negative consequences. One has only to look back to recent history to discover that many metropolitan areas within these states experienced substantial declines in employment at the end of the Vietnam war.

7. The current increase in defense spending has begun to trickle down to communities in many areas of the country. Boosters of defense spending caution, however, that such increases may crowd out commercial investment and fail to meet commercial goals of industry (Bruckner, 1984).

8. Portions of this project were completed under contract for the Office of Technology Assessment, contract 333-670.0. The development of the data base and the majority of the descriptive analysis were done under contract with the National Science Foundation, contract SES 82-08104.

REFERENCES

ARMINGTON, C., C. HARRIS, and M. ODLE (1983) Formation and Growth in High-Technology Businesses: A Regional Assessment. Washington, DC: Brookings Institution, Business Micro Data Project.

BOLTON, R. (1966) Defense Purchases and Regional Growth. Washington, DC: Brookings Institution.

BROWN, L. (1983) "Can high-technology industries save the Great Lakes?" New England Economic Review (November/December): 19-33.

BRUCKNER, L. (1984) The VSIC Program. Berkeley: University of California, Roundtable on the International Economy. (unpublished).

CHINITZ, B. (1961) "Contrasts in agglomeration: New England and Pittsburgh." American Economic Review Papers and Proceedings 51: 279-289.

DEUTERMAN, E. (1966) "Seeding science-based industry." Business Review: 7-15.

Fantas Corporation (1983) "Local high technology initiatives study." Reported to the Congressional Office of Technology Assessment, U.S. Congress.

Florida Department of Commerce (1982) "Study of the growth of employment in high-technology industries in the Southeast." (unpublished)

GLASMEIER, A. (1985) "Spatial differentiation of high technology industries: implications for planning." Ph.D. dissertation, University of Calfornia, Berkeley.

GLASMEIER, A., P. HALL, and A. MARKUSEN (1984) "Can everyone have a slice of the high-tech pie?" Berkeley Planning Journal 1 (Summer): 131-142.

——(1983) Recent Evidence on the Spatial Tendencies of High Technology Industries: A Preliminary Investigation; Working Paper 417. Berkeley:

University of California, Institute of Urban and Regional Development.

GLASMEIER, A., A. MARKUSEN, and P. HALL (1983a) Defining High-Technology Industries: Working Paper 407. Berkeley: University of California, Institute of Urban nd Regional Development.

——(1983b) Estimating Employment for Four- Digit High-Technology Industries: Working Paper 411. Berkeley: University of California, Institute of Urban and Regional Development.

GREEN, R., P. HARRINGTON, and R. VINSON (1983) High-Technology Industry: Identifying and Tracking an Emerging Source of Employment Strength. Boston: Northeastern University, Center for Labor Market Studies.

HALL, P. (1982) "Innovation: key to regional growth." Transaction/Society 19(5): 48-52.

HARRIS, C. (1984) "High-technology entrepreneurship in metropolitan areas." Washington, DC: Brookings Institution. (unpublished)

ISARD, W. (1960) Methods of Regional Analysis. Cambridge, MA: MIT.

KOCH, D., W. COX, D. STEINHAUSER, and P. WHIGHAM (1983) "High technology: the Southeast reaches out for growth industry." Economic Review (September): 4-19.

MALECKI, E. (1979) "Agglomeration and intra-firm linkage in R&D locations in the United States." Tijdschrift voor Economie en Society Geografie 70(6): 322-332.

MARKUSEN, A. (1984) Defense Spending and the Geography of High-Technology Industries: Working Paper 423. Berkeley: University of California, Institute of Urban and Regional Development.

National Governors' Conference (1983) State Activities to Encourage Technological Innovation: Preliminary Findings. Washington, DC: National Governors' Association.

NORTON, R. and J. REES (1979) "The product cycle and the spatial decentralization of American manufacturing." Regional Studies 13: 141-151.

Office of Technology Assessment, Congress of the United States (1983) Census of State Government Initiatives for High-Technology Development: Background Paper. Washington, DC: U.S. Government Printing Office.

PELTZ, M. and M. WEISS (1984) "State and local government initiatives for economic development through technological innovation." Journal of the American Planning Association 50(3): 270-280.

PERLOFF, H., E. DUNN, Jr., E. LAMPARD, and R. MUTH (1960) Regions, Resources, and Economic Growth: Resources for the Future. Baltimore: Johns Hopkins University Press.

REES, J. (1979) "Technological change and regional shifts in American Manufacturing." Professional Geographer 31(1): 45-54.

RICHIE, R., D. HECKER, and J. BURGAN (1983) "High technology today and tomorrow: a small slice of the employment pie." Monthly Labor Review (November): 50-58.

RONES, P. (1980) "Moving to the sun: regional job growth, 1968-1978." Monthly Labor Review (February): 12-19.

SAXENIAN, A. (1981) Silicon Chips and Spatial Structure: The Urban Development of the Santa Clara Valley; Working Paper 345. Berkeley: University of California, Institute of Urban and Regional Development.

STANBECK, T. and T. NOYELLE (1984) The Transformation of American Cities. Totowa, NJ: Rowan and Allenheld.

Technical Marketing Associates (1979) High Technology Enterprise in Massachusetts: Its Role and Its Concerns. Boston: Author.

U.S. Department of Commerce (1983) United States Industrial Outlook. Washington, DC: U.S. Government Printing Office.

VINSON, R. and P. HARRINGTON (1979) Defining "High-Technology" Industries. Boston: Massachusetts Department of Manpower Development.

VINSON, R., R.L. GREEN, and P. HARRINGTON (1983) High Technology Industries: Identifying and Tracking an Emerging Source of Employment. Boston: Northeastern University, Center for Labor Market Studies.

WEINSTEIN, B. and R. FIRESTINE (1978) Regional Growth and Decline in the United States. New York: Praeger.

4

Silicon Valley and Route 128:
Regional Prototypes or Historic Exceptions?

ANNALEE SAXENIAN

□ SILICON VALLEY has captured the public imagination. This California region has become a symbol of the benefits of microelectronics-led development. To most it represents the innovative dynamism that "high tech" will bring to America's lagging economy. And to policy makers in depressed regions across the country, it is a model to study, one they hope to replicate.[1] Yet few are aware of the contradictory nature of the Silicon Valley experience. Although microelectronics brought explosive growth to the Santa Clara Valley, it brought a panoply of social problems in its wake. Skyrocketing housing costs, severely congested freeways, and dangerous levels of air and water pollution now plague local residents and manufacturers. By the late 1970s concerned citizens had organized a vocal no-growth movement demanding a halt to further industrial development, and local companies had formed a manufacturer's association to deal with the declining "overall attractiveness of Santa Clara County as a place to live and work" (Santa Clara County Manufacturing Group, 1979). Faced with labor shortages and escalating costs, the county's largest employers have stopped expanding operations in the valley, choosing to disperse their new production operations to distant locations.

To what extent can the experience of Silicon Valley be seen as a prototype for the evolution of other regional concentrations of microelectronics production? This chapter explores these issues in a comparative context. The first section is a detailed examination of the interactions between the evolution of the semiconductor industry and the development of the region and its urban geography.[2] Parallels with the transformation of the Route 128 region in Massachusetts—the East Coast counterpart of Silicon Valley—are then highlighted in the second section. Not only did analogous circumstances condition the rapid postwar growth of these two regions, but their subsequent social and urban evolutions show striking similarities. The concluding section argues that the urban problems of Silicon Valley and Route 128 are rooted in the social structure generated by science-based industry. Although replication of these regional seedbeds of technological innovation is unlikely, their urban landscapes represent a scenario for the future to those communities that attract high tech production, and a challenge for planners, policy makers, and residents of these regions.

THE SILICON VALLEY STORY

The development of Silicon Valley has been inextricably linked to the evolution of the microelectronics industry.[3] Both trace their origins to World War II.[4] The electronics-related companies that established operations in the area during the 1940s and 1950s paved the way for the location of the valley's first semiconductor companies. The subsequent agglomeration and expansion of this new industry provided the impetus for its accelerated growth.[5] When the market for microelectronics boomed in the late 1960s and 1970s, so did the region. By 1970 this former agricultural valley had become one of the wealthiest and fastest growing urban areas in the nation. A decade later the region was also distinguished by urban and social problems that threatened the operations of the very companies that had brought its explosive growth.

Silicon Valley provides a rare opportunity to examine the links between an industrial sector, a region's growth, and its urban spatial structure. Prior to 1940 Santa Clara County—which encompasses a valley flanked by low coastal mountains—was blanketed with fruit orchards and farms. Today the microelectronics industry dominates the region and few traces remain of its agricultural past. Over 70% of the county's manufacturing workforce is currently employed in high tech sectors, while many of the remainder are employed in occupations that service or support this high technology complex (see Appendix). Most of the region's postwar growth is thus an outcome of the growth of this sector. Similarly, as there was little prior urban development, the current organization of space in the county can be traced to the characteristics of production in the industry. These relations between production, regional growth, and urban geography occur everywhere; but it is unusual to find a case in which spatial and regional outcomes are so clearly attributable to a particular sector.

The scientific base of the semiconductor industry distinguishes it from traditional manufacturing industries in many ways. Most salient is the occupational structure that it generates. The production process requires large numbers of highly educated engineers, scientists, and other technical professionals, in addition to the managerial strata and the production workforce. In 1972, for example, 40% of the U.S. semiconductor industry's workforce were nonproduction employees—mainly professional and technical—compared with only 16% in the production of motor vehicles and 13% in the apparel industry (Mutlu, 1979). An equal proportion of the workforce was employed in unskilled and semiskilled production and maintenance occupations.[6] The industry thus employs an unusually large proportion of professional and managerial employees alongside an equally large, but minimally skilled, production workforce. This top-heavy and bifurcated labor force fundamentally shaped the social structure and urban geography of Silicon Valley.

Stanford University provided the focal point for the innovative activities and new firm start-ups in Santa Clara County during the 1950s and 1960s. The region's first microelectronics enterprises were located in Palo Alto. For these technologically sophisticated but experimental ventures—many of which were started by Ph.D. scientists and engineers leaving academia for the first time—the university town provided a familiar and supportive intellectual environment. This pattern was consolidated with the establishment of the Stanford Industrial Park. Located on 770 acres adjoining the Stanford campus, it was devoted solely to high technology enterprises. Entrepreneurs who were unable to find space in the industrial park chose to locate in the surrounding area in order to be "close to the action" in a highly competitive and rapidly changing industry. The adjacent towns of Mountain View and Sunnyvale, and later Cupertino and Santa Clara, quickly recognized the tax revenue benefits of a strong industrial base. Following the Stanford model, they established industrial parks and provided various financial and infrastructural incentives to attract technology-based companies.

Industrial development continued to cluster in the northwest corner of the Santa Clara valley. Over time this produced a striking imbalance in the county's land use. Electronics production is now concentrated primarily in the cities of the north county, the "jobs belt" (see Figure 4.1). By 1970 Palo Alto-based firms provided 1 electronics job for every 4 city residents, while 20 miles to the south,

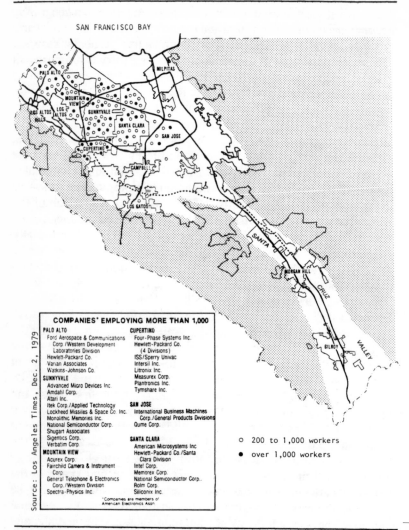

SAN FRANCISCO BAY

Source: Los Angeles Times, Dec. 2, 1979

COMPANIES' EMPLOYING MORE THAN 1,000

PALO ALTO
Ford Aerospace & Communications
Corp./Western Development
Laboratories Division
Hewlett-Packard Co.
Varian Associates
Watkins-Johnson Co.

SUNNYVALE
Advanced Micro Devices Inc.
Amdahl Corp.
Atari Inc.
Itek Corp./Applied Technology
Lockheed Missiles & Space Co. Inc.
Monolithic Memories Inc.
National Semiconductor Corp.
Shugart Associates
Sigentics Corp.
Verbatim Corp.

MOUNTAIN VIEW
Acurex Corp.
Fairchild Camera & Instrument
Corp.
General Telephone & Electronics
Corp./Western Division
Spectra-Physics Inc.

CUPERTINO
Four-Phase Systems Inc.
Hewlett-Packard Co.
(4 Divisions)
ISS/Sperry Univac
Intersil Inc.
Litronix Inc.
Measurex Corp.
Plantronics Inc.
Tymshare Inc.

SAN JOSE
International Business Machines
Corp./General Products Divisions
Qume Corp.

SANTA CLARA
American Microsystems Inc.
Hewlett-Packard Co./Santa
Clara Division
Intel Corp.
Memorex Corp.
National Semiconductor Corp.
Rolm Corp.
Siliconix Inc.

'Companies are members of
American Electronics Assn.

○   200 to 1,000 workers
●   over 1,000 workers

Figure 4.1   Location of Electronics Employment in Silicon Valley, 1979

in the city of San Jose, there was 1 electronics job for every 50 residents (Keller, 1979).

As industry and jobs flowed into the north county, the southern city of San Jose followed a different route. The city's administration, supported by a coalition of landowners, realtors, contractors, road builders, speculators, and bankers, was explicitly committed to "making San Jose the Los Angeles of the North." They promoted rapid urban expansion by rezoning large tracts of land, aggressively annexing nearby territory, and extending sewer systems, storm drains, and roads to peripheral areas. Local developers in turn availed themselves of FHA mortgage financing for the tract development of inexpensive, single family homes. San Jose thus grew from an agricultural processing and distribution center of only 17 square miles in 1950 to a sprawling 147-mile "suburb" in 1975. Today close to half of the total population of Santa Clara County lives in this "bedroom city."

The explosive growth of Silicon Valley's microelectronics firms was quickly reflected in the local labor markets.[7] Total employment in the county doubled during each successive postwar decade, resulting in the creation of more than 500,000 new jobs between 1940 and 1975. A massive wave of immigration in turn provided the labor pool needed to meet the growing demands of local companies. The region's population thus increased by more than one million during this period, with natural increases accounting for only one-quarter of annual population growth.

Coming in a dual stream, Silicon Valley's immigrants reflected the microelectronics industry's labor requirements. On one hand, an influx of professionals and highly educated workers responded to the industry's demand for scientists and engineers. Close to half of the adult immigrants to the region during the 1950s and 1960s had some college training. By 1970 40% of the county's population was college educated, compared to only 20% in 1940. At the same time, the industry's demand for production workers attracted large numbers of unskilled, predominantly minority individuals. These included displaced agricultural workers from California and the Southwest (primarily Latinos and some Philippino Americans), foreign born Mexicans and Philippinos, and a smaller number of U.S. Blacks, and native Americans. By 1970 these minority groups together represented one-quarter of the county's total population.

The settlement of this dual stream of immigrants into socially and economically segregated residential communities shaped the urban geography of Santa Clara County. While engineers and professionals tended to settle in the north of the county, in Palo Alto, and in the foothills to the west, most of the less skilled immigrants landed in the sprawling southern city of San Jose. For analytic purposes, the county's 15 cities can be divided into 4 homogeneous spatial clusters, each representing a specific social and/or economic function in the county.[8] (These regions are delineated in Figure 4.2).

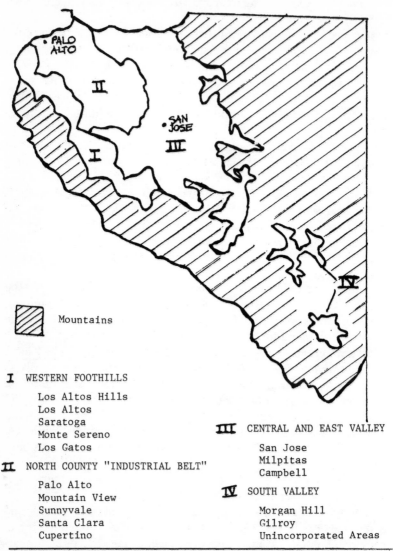

Figure 4.2   Social-Residential Segregation in Silicon Valley

Santa Clara County's most affluent professionals and executives reside in the western foothills (Region I). The cities in this cluster are solely residential communities, and they are the newest in the county. They were settled during the 1950s by the influx of entrepreneurs and scientists who came to work in the emerging microelectronics industry, and who desired both the spacious beauty of the foothills and the proximity that the location provided to their new workplaces. The residents of these foothill cities earn incomes that dwarf those in the rest of the valley, they have the highest level of educational attainment in the county, the majority are in professional and managerial occupations, and virtually all are white.

The 5 cities of the north county (Region II) form the heart of the Silicon Valley technology complex. Their robust industrial base provides those 5 cities alone with half the county's jobs and the highest municipal property values in the region. The residents of these north county cities represent an intermediate socioeconimic mix. Palo Alto and the far western city of Cupertino are clear counterparts of the affluent foothill cities in terms of incomes, education levels, occupational status, and racial mix. Moving southward from Palo Alto, city-by-city, the socioeconomic status of the population steadily declines, the occupational mix is dominated progressively by craftsmen and operatives, and the minority population rises dramatically.

This trend culminates in the older cities of the central and east valley (Region III), where over half of the county's population resides. Dominated by San Jose, these cities served as a landing spot for the influx of unskilled, often minority inmigrants who were attracted by the electronics boom. With incomes below the county average the residents of these central and east valley "bedroom communities" are less well-educated and more likely to be of minority descent than their neighbors to the north and west.

The cities of the south valley and the unincorporated areas (Region IV) are the home of the poorest 10% of the county's population and a disproportionately high percentage of its minority residents. Until recently they remained rural and agricultural, with minimal connections with the industrial complex.

Silicon Valley's urban landscape has thus come to mimic the microelectronics industry's bifurcated social structure. The large and affluent professional-managerial strata of the industry's workforce is insulated in the north and west—in Palo Alto and, to a lesser extent, in the other north county industrial cities—with easy access to the electronics complex. Meanwhile, the industry's large, low-income production workforce is concentrated further to the south and east, primarily in San Jose and the adjacent communities. This social-spatial segregation is highlighted by comparison of a few statistics. In 1970 50% of Palo Alto's employed residents were in professional, technical, and managerial occupations and 42% of the city's adults had four or more years of college education, whereas only 29% of San Jose's employed residents were in professional, technical, and managerial occupations and only 15% had college education. On the other hand, 25% of San Jose's employed residents were craftsmen or operatives and 22% of its population was of Spanish-American descent, while only 10% of Palo Alto's residents were craftsmen or operatives and under 6% were Spanish-Americans.

This urban form evolved in Santa Clara County during the 1950s and 1960s to accommodate the industry's bifurcated workforce within a single metropolitan area. Yet the limitations of this spatial structure soon began to manifest themselves. By the 1970s strains had appeared in the county's housing market, transportation networks, and environment. These urban problems in turn called forth the mobilization of local residents and manufacturers.

Silicon Valley's settlement patterns generated a chronic housing shortage and rapidly inflating housing prices. During the 1960s and 1970s the county's housing supply failed to grow in pace with its accelerated job growth. By 1980 there were over 670,000 jobs but only 480,000 housing units in Santa Clara County, and the average home was over $100,000—almost double the U.S. average (Santa Clara County Housing Task Force, 1977).

The housing shortage is a consequence of the restrictive land-use policies and exclusionary planning practices imposed by local governments in the north county seeking to preserve the spacious and rural nature of their affluent residential communities. In pursuit of expanded tax bases, these cities rezoned extensive tracts of land from residential to industrial usage. Between 1965 and 1975 alone, such rezoning reduced the countywide housing capacity by 54%, a loss of 417,000 potential housing units (Bernstein et al., 1977). Low density zoning and residential growth management policies have further limited the county's potential residential land. In 1972, for example, Palo Alto zoned all of its foothill areas exclusively for open space; and in the western foothills virtually all residential construction is restricted to single- rather than multiple-unit housing. Today there is, on average, only one dwelling unit per acre of residential land in these communities. In the very exclusive city of Los Altos Hills, there are two acres for every home (Santa Clara County Manufacturing Group, 1980).

Within this context, the microelectronics industry's top-heavy employment structure directly fuels housing price inflation. Any industrial expansion almost invariably draws significant numbers of highly paid engineers, scientists, and managers into the county. With incomes that are 5 to 10 times those of the industry's production workers, these newcomers have asserted their superior buying power in competition for the valley's scarce housing, bidding prices up to levels that only their incomes can bear and often displacing low-income residents in the process.[9] By 1980 a typical middle-class home in the valley was priced over $150,000, more than double that of a comparable home in such western states as Texas and Arizona.[10] The same house in Palo Alto was priced at close to $300,000.

This process has in turn exacerbated the county's transportation problems. The sprawling nature of postwar residential development and the spatial imbalance of jobs and housing location have always meant long commutes for some of Santa Clara County's residents. With no viable mass transit alternatives—municipal buses typically take twice the driving time—local residents are dependent on private automobiles for transportation. Today, as the least costly homes and apartments are increasingly found only in the southern- and eastern-most reaches of the county, the industry's production workers are being forced to locate (or relocate) further and further from the jobs belt in the north—often up to 50 miles from their workplaces.

By 1975 over 4 million automobile trips were being taken daily in the county. Of those, the majority were home-based work trips, and the great distances separating jobs and housing meant that most of these commutes were long, slow, and costly. County traffic engineers now report severe rush-hour congestion on all of the county's major freeways and expressways when travelling from the southeast to the northwest during the morning peak hour—from housing to jobs—and in the reverse direction during the evening rush. Workers living in the south county now face commutes of three hours or more daily.

The negative effects of congestion and over-burdened traffic networks fall most heavily on those who commute from the south and east—primarily the lower-income workers who can least afford the rising fuel costs. However, all commuters now suffer delays and frustration. The overflow of traffic onto residential streets has brought noise, accidents, and disruptions to even the most affluent north county neighborhoods. And none can escape the degradation of the region's air quality, which has resulted from excessive automobile use. Even living in the isolated foothill communities is no longer a protection from these urban disamenities.

Auto emissions have generated the brownish-orange clouds of smog that now hang over the valley, where one or more federal air quality standards are violated at least 10% of the time (Santa Clara County Industry Housing Management Task Force, 1979). Meanwhile, the severely limited tax bases in the central and east valley cities have resulted in chronic revenue shortfalls. The recent breakdowns of San Jose's sewage treatment plants—which have seriously damaged the aquatic life and water quality in San Francisco Bay—are one consequence of the city's inability to finance adequate upkeep of sewers, parks, roads, and other basic public works. The fruit orchards and agricultural fields of the past have disappeared, making way for the almost ubiquitous strip development of shopping centers, industrial parks, and freeways. In 1971 the city of San Jose maintained only 8 acres of open space per 1000 residents, compared to 35 acres in nearby San Francisco and 25 in New York City (Stanford Environmental Law Society, 1971).

The environmental and social amenities that once distinguished Santa Clara County are thus rapidly being undermined. In 1974 an observer described the attractions of Silicon Valley for the industry's professionals as follows:

> It's a particularly pleasant place to live and work, a beautiful landscape of hills and plains, a bounteous garden of nature where fruit trees and wild flowers bloom even in February. . . . Few places on earth so agreeably mix hedonistic delights with the excitement of urbanity. Outdoor sports and recreation are year round attractions. . . . The area boasts 4,000 PhDs. There are at least 12,000 horses, some kept by those PhDs right on their home acreages, which are often within minutes of work (Bylinsky, 1974).

These qualities—which have made the valley especially desirable for the microelectronics industry's executives and scientists—are now being threatened. At the same time that housing prices are becoming prohibitive, living in the isolated foothill communities no longer guarantees an escape from the region's urban woes.

Silicon Valley's no-growth movement is a clear reaction to the declining quality of life in the county. An alliance of local environmentalists, planners, housing activists, women's groups, elected officials, and neighborhood groups, this coalition emerged in the 1970s on the momentum of several countywide task forces that had examined the region's housing, transportation, and environmental problems. In their view, uncontrolled job growth is the primary cause of the region's urban and environmental problems. Their position is represented in the report by the county's Industry and Housing Management Task Force, *Living Within Our Limits* (1979), which advocates the imposition of strict governmental controls over the rate, amount, type, and location of further industrial development. In January 1980 the city of Sunnyvale shocked the rest of the valley by imposing a 4-month moratorium on all new industrial projects.

Santa Clara County thus became the first area in the country to legislate public controls over industrial growth. Although the task forces' proposals for countywide planning of job growth and revenue sharing were defeated at the regional level, most of the governments of the north county cities enacted municipal controls on development. The Sunnyvale city council legislated strict regulations on the employee density in local plants, imposed fees on all new industrial construction in order to finance improvements in the transportation system, and rezoned substantial amounts of land from industrial to residential usage. Similarly, Palo Alto now requires all developers of new industrial projects to contribute to a fund for low- and moderate-income housing in the city. These cities have also imposed strict regulations concerning the environmental impacts and physical appearance of new development.

Silicon Valley's high technology companies soon began to feel the strains as well. By the 1970s even the most prestigious microelectronics companies were unable to attract experienced engineers and scientists from outside of the region because of the inflated housing costs.[11] A 1979 survey of 68 local firms showed the salaries of engineers in the valley rising an average of 15% to 20% a year. These companies also suffered from a shortage of production workers and unusually high turnover rates as affordable housing disappeared and the workers began rejecting the long and expensive commutes to work. As the personnel manager of one firm noted, "We're in trouble with our commuting patterns. Eventually the local labor force isn't going to be able to get here because of the crowded highways." In 1980 there were over 10,000 unfilled jobs of all skill levels in Santa Clara County (SRI International, 1980).

After almost 30 years of operating in an explicitly progrowth environment, with virtually no dealings with local governments, Silicon Valley's industrialists also entered the political arena. In 1978 51 of the county's largest employers formed the Santa Clara County Manufacturing Group. While most of its members already belonged to industry associations such as the Semiconductor Industry Association and the American Electronics Association, this group was formed to deal explicitly with regional issues. Their official mandate is to organize local industry to "cooperate in addressing major issues affecting the overall attractiveness of Santa Clara County as a place to live and work." They were active in the debate during 1980 over growth controls and development restrictions. In addition to fighting the imposition of limits to industrial development, they participated in framing the public debate and in conducting their own studies and devising solutions for local housing, transportation, and environmental problems.

The activities of the Manufacturing Group are an indication of the industry's dependence on the region. These firms recognize the need to preserve the environmental, social, and cultural amenities of the valley. Only a desirable quality of life will ensure their ability to attract and hold the experienced engineers and professionals who are so critical to their continued growth.

Relocation, however, has gone hand in hand with political mobilization for many of these companies. Departing from past policies of expanding exclusively in the valley, virtually all of the county's large microelectronics companies have begun building in distant locations. A spokesman for American Microsystems, Inc., explained their first decision to locate outside the region: "Housing costs here are outlandish compared to elsewhere. You can get an awfully good home for $85,000 in Pocatello (Idaho). . . . The cost of living is 30 to 40% cheaper." Like AMI, which discovered the virtues of Pocatello, Intel built a new manufacturing facility in New Mexico and expanded its operations in Oregon and Utah. National Semiconductor turned a 55-acre plot in Santa Cruz into a park for its employees, while building new plants in Arizon and Utah. Advanced Micro Devices and Memorex both opened new factories in Texas, Signetics chose Utah, and Hewlett-Packard expanded in Oregon, Colorado, and Washington. These companies do not intend to expand their manufacturing activities in Silicon Valley in the future.

This decentralization of production is a response to labor shortages and escalating costs in the region. However, the concommitant consolidation and restructuring of the industry itself has enabled these shifts. By the late 1970s Silicon Valley's major chip manufacturers had changed from small, intensely competitive, technology-dominated ventures to large, mature, marketing-oriented corporations.[12] The five leading microelectronics companies in the region—all of which rank among the world's top ten—now dwarf the rest of the county's producers in terms of both sales and employment. Increased scale has given them the financial ability to disperse their manufacturing operations and, in conjunction with the changing nature of competition in the industry, freed them from the need to agglomerate in Silicon Valley.

Yet the industry is not leaving the valley altogether. Hewlett-Packard recently completed a new corporate headquarters building in

Palo Alto and AMD began operations in its new Technology Development Center in Sunnyvale. As manufacturing growth is directed elsewhere, a new interregional division of labor is evolving in the microelectronics industry.[13] Silicon Valley is gradually being transformed into a high-level research and control center—the site of headquarters and sophisticated research, product design, development, and prototype production activities for the major microelectronics corporations, as well as a site for the start of new technology-based companies. Meanwhile, regions in the South and West are growing as manufacturing centers for the industry's standardized products.[14]

THE ROUTE 128 REGION

The winding country road that served as a north-south bypass around downtown Boston was transformed into a four-lane superhighway in 1951. Completion of the Boston Circumferential Highway—known today as Route 128—coincided with the ballooning of the region's advanced technology industries. During the succeeding decades the Route 128 region rapidly ascended to its current position as the East Coast counterpart of Silicon Valley's technological-industrial complex.[15] Yet, as in Silicon Valley, by the late 1970s the towns in the Route 128 region were beset by urban, environmental, and social problems; and local firms were beginning to locate their new operations outside of the "golden semicircle."

In 1955 there were 40 companies settled along the golden semicircle. Ten years later there were close to 600 companies on the Route 128 roster, including servicing and distribution firms as well as defense plants and other technology-based companies. By the mid-1970s CC&F had built 16 industrial parks along Route 128; and, faced with congestion and overcrowding, it had also initiated the development process along Route 495, another circumferential highway located about 15 miles out from Route 128 (Adams, 1977).

While the historic conditions surrounding the development of a technological complex in the Route 128 region are comparable to those in Silicon Valley, documenting parallels to the evolution of its urban geography is more difficult. There are two problems: First, the very nature of a circumferential highway that disperses development through portions of many different governmental jurisdictions makes data collection and comparison difficult.[16] Furthermore, in the absence of physical constraints, such as mountains, the appropriate boundaries for the region are difficult to define. Many Route 128 firms draw on labor markets outside of Massachusetts altogether, with workers commuting from Maine, New Hampshire, and even Rhode Island. Second, technology-based industries are not the region's dominant employers as they are in Silicon Valley. In 1980 high tech activities accounted for 31.6% of manufacturing employment and only 8.5% of total employment in the region. Even in absolute terms, high tech industry employs 30% fewer employees in the Route 128 region than it does in Silicon Valley (see Appendix). Furthermore, the area has a long industrial history of textile and other mill-based production that shaped both the inherited urban landscape and the characteristics of the local labor force. Thus, urban and social outcomes cannot be so directly attributed to the growth of this sector as in the case of Silicon Valley.

Despite these differences, certain parallels are evident. Rather than spreading evenly along the highway, industrial development has remained highly concentrated in select cities along Route 128. In addition, a distinct pattern of social-residential segregation has evolved in its suburbs. Today the region is experiencing major social and planning problems. Many of its roadways are seriously overcrowded, housing prices have inflated, and its research-based firms are experiencing labor shortages. Local citizens, concerned about the destruction of the environment and the declining quality of life, have organized movements to limit industrial growth in their communities. Meanwhile, the largest technology-based firms have begun to locate their new operations outside of the golden semicircle; and since 1976, when local industry executives formed the Massachusetts High Technology Council, they have become a major lobbying force in state affairs.

Route 128 passes through 20 different towns, each with distinctive zoning ordinances (see Figure 4.3). These variations have shaped the location of industry and of various classes of residential development in the region. Burlington and Waltham, for example, which put virtually no restrictions on industrial development, have been flooded with high tech industry. Bedford, which is especially attractive to electronics companies because of the location of MIT's labs at its Hanscom Field Air Force base, has carefully confined the town's industrial growth to two discrete locations. Lexington, on the other hand, allows virtually no industry, although it is more receptive to pure research laboratories.

Residential settlements in the area, which expanded rapidly during the 1950s and 1960s to serve the Route 128 complex, are also highly

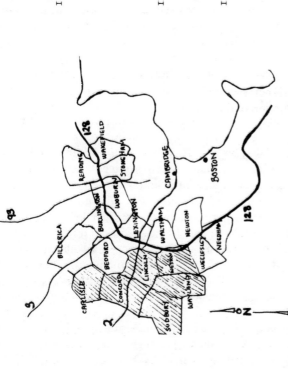

	Median Value of Owner-Occupied Housing
I. HIGH TECH TOWNS	
Wakefield	$ 61,500
Reading	65,200
Stoneham	60,000
Woburn	52,600
Burlington	61,800
Billerica	52,400
Bedford	75,300
Waltham	59,500
II. ROUTE 128 SUBURBS	
Lexington	85,200
Newton	81,200
Wellesley	99,400
Needham	78,200
III. EXCLUSIVE SUBURBS	
Lincoln	124,700
Carlisle	99,000
Concord	89,000
Sudbury	90,000
Wayland	85,900
Weston	143,600

Communities with minimum lot size of 40,000 sq. ft. or more zoned for over 75% of total residential area.

SOURCE: U.S. Bureau of the Census, 1980 Census of Population and Housing.

Figure 4.3 Residential Suburbs in the Route 128 Region: Median Housing Values, 1980

differentiated. The towns on or immediately adjacent to Route 128 fall into two categories. There are the decidely proindustry, high tech boom towns, such as Burlington, Woburn, and Billerica, that boast loose zoning, low property values, and blue-collar populations. These heavily industrialized towns are clustered primarily along the northern stretch of Route 128, with access to north-bound Routes 3 and 93. Waltham, located further to the south, also belongs in this industrial cluster; while Bedford, with its conscious planning of industrial and residential growth, provides a bridge to the other category of Route 128 towns. These moderately wealthy, more conservative suburban towns, such as Lexington, Newton, and Wellesley, have significantly higher property values and strict limits on industrial development.

A cluster of highly exclusive residential suburbs is located along Route 2, extending westward from Route 128. Route 2 crosses Route 128 near its midpoint and feeds directly into Cambridge and Boston along the Charles River, passing MIT and Harvard en route. It thus provides a perfect access road for the engineers, scientists, and other professionals who live in towns like Concord and Lincoln and work either in Cambridge or on Route 128 (or both). This cluster of towns is distinguished by severely restrictive residential zoning regulations and, correspondingly, the highest housing prices in the Route 128 region. One observer has noted the social status associated with living in these towns: "These are quiet, leafy places with good public schools, and scientists with young children like to live in them" (Rand, 1964).

By the late 1970s, as the region's technological industries continued to grow, severe urban and environmental strains began to appear along Route 128. Take, for example, the case of Burlington. Burlington is located at the confluence of a number of major transportation routes—where Route 128 interchanges with Routes 3 and 3A. The desirability of this location, in combination with an extremely conducive zoning code, attracted an initial influx of commercial and industrial developers during the 1950s. This industrial growth became self-reinforcing and, with over half of the city's land allocated to nonresidential uses, it peaked during the late 1960s and 1970s.

Burlington was quickly transformed from a sparsley populated, rural town into a hub of commerce and high technology industry. The city's population increased over 600% between 1950 and 1970— exhausting virtually all of the land available for residential development (Metropolitan Area Planning Council, 1977). As population growth drew to a halt, however, commercial and industrial construction accelerated. Employment in the city more than doubled during the 1970s (Hilgenhurst et al., 1983). The result was a jobs-housing imbalance. The local Chamber of Commerce claims that by 1980 there were only 9,600 employable residents and over 30,000 jobs in Burlington (Boucher and Denison, 1982). According to another estimate, only 10% of the workers in the city's high technology firms, shopping malls, and office buildings actually live inside the city. The remaining 90% commute (Boston Globe, 11/20/82).

The same high tech development that brought jobs, prestige, and financial solvency also brought severe traffic problems and pollution to Burlington. Rush-hour traffic is so congested along the city's roadways that commuters spend 30 minutes driving 8 blocks across town to Route 128 or else travel miles out of their way to avoid the congestion. Pollution is also a concern for Burlington residents. Several municipal wells have been closed down in recent years due to contamination, and traces of triclorethylene—a toxic cleaning fluid used by high tech companies—have been discovered recently in the town's water supply. Finally, in early 1982 the town's engineers discovered that chemical solvents from industrial wastes had eroded large sections of a sewer main, which was replaced at a cost of $300,000 (Boucher and Denison, 1982).

In 1982 a powerful grass-roots movement coalesced in Burlington, demanding a halt to further industrial and commercial expansion. A petition was presented at a town meeting, which resulted in a 3-month moratorium on construction in 3 heavily developed areas of the city and the formation of a committee to recommend controls on future development. The citizen's organization—The Middlesex Turnpike Improvement Association (MTIA)—has called public attention to the link between explosive industrial development and "the threat to our water supply, the increase in accidents and crime and pollution which affect everyone in the community" (1982). Although the moratorium ended without significant new restrictions being placed on development, the MTIA has become a vocal force in local politics and is now fighting its battle electorally, pushing for the replacement of prodevelopment politicians by no-growth candidates.

These experiences have been repeated in many Route 128 towns and cities. Despite success in restricting industrial development to two isolated locations, the traffic problems in the town of Bedford are virtually insoluble since, according to a town planner, "we have about twice as many people working here as live here," and local roads are carrying 10 times as many cars as they were designed to handle (Boucher and Denison, 1982). Four of the town's eight wells were shut down in 1979 due to contamination, and the sewage system is now seriously overloaded. Faced with the expenses of accumulating infrastructure repairs—which are beginning to outweigh the industry's tax contributions—as well as the declining quality of life, a large group of Bedford residents is now calling for a halt to high tech development.

Route 128's technology-based companies face a related set of problems. The president of Digital Equipment Corporation has explained that the firm intends to locate all new plants outside of Massachusetts because, "I don't think the roads in the area can take any more people" (Boston Globe, 11/5/82). Meanwhile, Q.S.C. Industry recently expanded in New Hampshire because, "the local labor base was exhausted. There was just no way to expand there and attract good people" (New York Times, 10/20/80). As a result, the region's largest companies have already completed several out-of-state expansions. Digital Equipment Corporation, for example, opened new plants in Phoenix, Albuquerque, Colorado Springs, and Greenville, South Carolina. Data General made the move to Austin, Raleigh, North Carolina, and Portland, Maine. Smaller companies such as Teradyne and Analog Devices are also considering such moves.

At the same time that Route 128 firms are beginning to consider out-of-state expansions, the industry's executives have banded together to form the Massachusetts High Technology Council. This industry association, formed in 1977 to represent over 100 of the state's electronics and related firms, has become a dominant force in Massachusetts politics. The group has pursued a political agenda of major tax and budget cuts on the grounds that the availability of skilled workers—the key to the future prosperity of the industry (and, thus, the region)—is threatened. Member firms report increasing difficulties in recruiting senior-level scientists and experienced managers and engineers because of the region's high cost of living relative to competing states. A 1980 study, for example, showed that the total cost of transportation, housing, and taxes for an average engineer was 50% higher in a Route 128 suburb than in such growing technology centers as Austin, Texas and North Carolina's Research Triangle Park (Business Week, 3/10/80). Their campaign to cut state and local personnel taxes is thus intended to eliminate what they claim to be a major obstacle to recruiting experienced engineers from out of state. This explains the substantial economic support that the High Tech Council provided for the passage of Proposition 2½, a 1980 measure that mandated severe reductions in local property taxes in Massachusetts.

Just as no-growth movements represent the attempts by local residents to preserve the quality of life in their communities, this mobilization of local industry is aimed at maintaining the viability of the Route 128 region as a location for high tech activity. Although manufacturing branch plants are now being located outside of the region, headquarters and research and development activities remain, along with the numerous start-ups and smaller technology-based firms that characterize the region. These phases of production are precisely those that require the highest quality scientists and the most experienced enginers. Faced with the rising cost of living in the region, the High Tech Council is attempting to preserve the region's ability to attract and retain this segment of the labor force.

CONCLUSION

Silicon Valley and Route 128 both grew out of a historically unique confluence of political, economic, and institutional circumstances. The origins of these seedbeds of technology-based industry lie in World War II, in the spending priorities of the Cold War, and in the development of close links between federal funding sources, local academic institutions, and local industry. It is unlikely that this particular combination of circumstances will be repeated. However, as the microelectronics and minicomputer industries have matured, their managers have learned to disperse rather than concentrate manufacturing growth. Thus, while such major centers of technological innovation will not be easily replicated, the patterns of urban development in Silicon Valley and Route 128 are not unique. Both regions illustrate the consequences of unplanned high tech development.

The agglomeration and expansion of microelectronics production transformed Santa Clara from an agricultural community into one of the fastest growing and most affluent regions in the nation within a few decades. It also generated an unusually top-heavy, bifurcated labor force and the imbalanced pattern of land use that now characterize Silicon Valley. Highly segregated suburbs evolved to accommodate the region's distinct social groups within the same metropolitan area. As the industry continued to grow, the limitations of this urban form became evident. Housing price inflation, transportation congestion, and labor shortages are physical manifestations of these limitations, while the mobilization of both residents and local industry are its social counterparts.

Evidence from the Route 128 region yields parallel observations. The evolution of its urban structure shows striking similarities to that of Silicon Valley: in particular, the spatial clustering of new industrial development alongside a hierarchy of socially segregated residential communities. In this case, severe transportation congestion is the most visible consequence of the growth of the region's high tech industry; while its infrastructural breakdowns and the rising cost of living are also rooted in this particular urban form. Similarly, the political responses—the emergence of both the no-growth activism and a regionally-based industry association—are clear counterparts to the Silicon Valley experience.

The limits to the growth of these regional concentrations of science-based industry are social, not physical. The environmental and cultural qualities that make a region attractive to engineers and other high tech professionals are subject to social scarcity. They are what Fred Hirsch refers to as "positional goods" (Hirsch, 1976). The availability of housing in a beautiful and isolated rural community with easy access to the cultural benefits of an urban center make both the Silicon Valley and the Route 128 regions attractive to engineers, scientists, and managers. However, these locations have been transformed by the very migrants they have attracted. Once moving to a secluded suburban community becomes possible for large numbers of people, the social qualities that make it desirable are undermined.

Strict zoning and planning restrictions, as in the most exclusive communities of both Silicon Valley and Route 128, are one means of minimizing the destructive side effects. These restrictions work, however, through an auction process that limits the number of newcomers. Once the limited supply of housing in such locations is filled, any growth of demand is reflected in rising prices (and current residents reap capital gains). As the science-based companies in these regions continued to recruit new engineers from out of state, they produced both shortages and rapidly inflating housing prices. The attempted solutions—outward expansion of suburban development to increase supply, social-residential segregation, and the spatial separation of industrial and residential development—have in turn fueled a different problem, traffic congestion.

Continued industrial growth in this context simply exacerbates a region's urban problems. Residents and firms doing business in both Silicon Valley and the Route 128 region have thus mobilized to counteract the self-destructive outcomes. The no-growth activists—the educated, professional, middle-class residents who desire to preserve the environmental amenities of the suburban communities in which they live—have called for a halt to industrial development; while local industrialists—faced with labor shortages and rising costs of production—have attempted to protect themselves from mutually competitive behavior by controlling the inflation of salaries and housing and transportation costs. Local firms also have an option that most residents lack: geographic relocation. The dispersal of manufacturing operations out of the region is a companion response to the same problems.

As manufacturing is increasingly dispersed from Silicon Valley and Route 128, new high tech regions are growing. Observers are already predicting the emergence of a "silicon desert" in Arizona, a "silicon prairie" in Texas, and a "silicon mountain" in Colorado. While it is highly unlikely that any one region will face an agglomeration of the sort that exists today in Santa Clara County or Massachusetts, similar urban problems are already evident in these regions. Labor shortages and transportation problems are being reported in high tech communities from Nashua, New Hampshire to Phoenix, Arizona. Or, for another example, take the city of Roseville, California. Hewlett-Packard, Signetics, and NEC Electronics have built there recently, and other electronics companies are planning growth in the area. A consortium of planning companies recently examined the effects of this growth on the region. They concluded that Roseville and surrounding towns have all zoned too much land for future industrial and commercial development. They also forecast a ten-fold increase in the region's population and severe housing and traffic problems.

In regions where science-based industries employ a large proportion of the local labor force, we can expect to see the emergence of such urban problems, themselves a product of the nature of production and the social structure generated by these industries. Over time the problems of social scarcity inherent in the nature of suburban housing, transportation facilities, and environmental resources will also result in the mobilization of both residents and local industrialists. The challenge for planners, businessmen and residents alike, is to avoid the pitfalls of the Silicon Valley and Route 128 experiences, and to plan for more rational urban development.

APPENDIX:
High Tech Employment:
Santa Clara County and the Route 128 Region, 1980

		Santa Clara County	Route 128 Region*
SIC 357	Office and computing machines	55,320	23,488
SIC 366	Communication equipment	19,603	33,619
SIC 367	Electronic components and accessories	59,925	18,870
SIC 376	Guided missiles, space vehicles, parts	10,000-25,000**	3,750
SIC 381	Engineering and scientific instruments	4,368	5,722
SIC 382	Measuring and controlling devices	16,488	14,618
SIC 383	Optical instruments and lenses	1,012	5,314
SIC 737	Computer and data processing services	7,813	13,850
SIC 7391	Research and development labs	3,856	7,402
Total high tech employment		185,885	126,633
Total manufacturing employment		256,437	400,685
Total employment		610,494	1,485,376

SOURCES: U.S. Department of Commerce, Bureau of the Census; County Business Patterns (1980); Massachusetts (CBP-80-23) and California (CBP-80-6).
NOTE: There is no commonly accepted definition of "high tech," or what to include in it. For a discussion of this problem see *Business Week* (3/30/83). The definition used here is that of the Association of Bay Area Government (ABAG) in California.
*Includes Essex, Middlesex, Norfolk, and Suffolk Counties.
**Midpoint is used in totals for categories in which range is given for disclosure purposes.

NOTES

1. The regional rivalry for science-based industry is not new. In the early 1960s most metropolitan areas with any university relationship were competing to make it onto the R&D map. Cleveland was building a research-oriented industrial park called University Circle, and similar projects were under discussion in university towns from Connecticut and Pennsylvania to Detroit, Chicago, and Dallas (Rand, 1964). Twenty years later, as older industrial sectors either stagnate or decline, the competition has reached a feverish pitch. According to a recent survey, 33 states have spent, or are planning to spend, a total of $250 million on R&D facilities to speed the growth of high technology industry within their borders (Wall Street Journal, 2/14/83); while thousands of local governments are scheming to devise an optimal package of tax breaks, incentives, and regulatory relief to lure these innovative firms to their turf. Attracting high tech has become the only development game of the 1980s.

2. The Silicon Valley section of this chapter is based on research conducted in 1979 and 1980 by Saxenian (1980).

3. The only major U.S. semiconductor firms that did not originate in Silicon Valley are Texas Instruments and Motorola, from Texas and Arizona respectively. IBM is the country's largest chip producer, but it does so only for in-house use, as does AT&T.

4. The location of hostilities in the Pacific laid the foundation for the industrialization of the agricultural Santa Clara Valley; while the federal funding of weapons research and development at Stanford University initiated the formation of a technological watershed in the region. The war also stimulated the emergence of the semiconductor industry. The invention of the solid state transistor—the industry's first product—at AT&T's Bell Labs was a response to military demand for small, versatile electronic components.

5. The history of the evolution of Silicon Valley has been told elsewhere. See, for example, Rogers and Larsen (1984).

6. While 40% of the industry's 1971 workforce were in managerial or professional positions, 48% were in production and maintenance occupations—four-fifths of which are classified as semiskilled or unskilled positions—with the remaining 12% in secretarial and clerical occupations. There is virtually no mobility between these sectors of the labor force (U.S. Dept. of Commerce, 1979).

7. The data for the following paragraphs is derived from a series of INFO fact sheets published by the Santa Clara County Planning Department in the years 1970 through 1980; along with their 1975 Special Countywide Census, 1978.

8. The analysis presented in this section is based on 1970 data. During the late 1970s some changes in this distribution of activity and people occurred, as high technology firms began to locate operations in San Jose and the south valley cities experienced accelerated population growth. These events have little bearing, however, on the sequence of events described here.

9. The average wage of a semiconductor production worker in 1979 was $4.52 an hour, or about $9,000 a year; while an experienced engineer currently earns over $100,000 a year. Top executives in the industry earn more than double that amount (U.S. Bureau of Labor Statistics, 1977).

10. According to the spring 1981 housing survey by the Nationwide Relocation Service, the average sale price of a comparable 3-bedroom, 2-bath house with a garage and family room was $292,500 in Palo Alto and $126,000 in San Jose, compared to $88,200 in Dallas; $72,000 in Phoenix, $81,000 in Boston, and $85,000 on Long Island (San Francisco Examiner, 7/5/81).

11. Not only are housing prices in the county 50% above other western states and close to double those in the rest of the United States, but their prime out-of-state competitors—Texas Instruments in Dallas and Motorola in Phoenix—are located in regions with substantially lower-priced housing.

12. This restructuring process was driven by the growing pressures of international competition, dramatic increases in the technological sophistication of semiconductor devices, the emergence of huge new markets which permitted the use of mass-production techniques and automation, and the precipitous rise in costs which have raised the barriers to entry dramatically. The late 1970s was thus a period of massive consolidation for the industry, as most of the small companies of the past were acquired or forced into mergers with larger electronic system producers or conglomerates. A fascinating and informative discussion of this process is contained in a five-part series in *Science* magazine (Robinson, 1980).

13. This is only one segment of the international division of labor in the industry. In the early 1960s virtually all of Silicon Valley's semiconductor companies relocated their most labor-intensive assembly operations to low-wage areas of the Third World, primarily Asia and Mexico. During the 1970s most major semiconductor firms also established manufacturing facilities in Europe in order to gain access to the markets of the European Common Market.

14. The spatial behavior of the semiconductor industry within the United States has thus conformed to the predictions of the product cycle theory of industrial investment pioneered by Raymond Vernon (1960, 1966). This theory postulates initial agglomeration in the development of a new product—allowing flexibility and the exploitation of the benefits of communication and external economies; followed by a large-scale dispersal of production facilities to lower cost locations as the product is standardized and demand for it grows—enabling the achievement of scale economies through mass output and long production runs. During this industrial maturation process, new technologies develop that allow the decomposition of production processes and their successive transfer outward.

Santa Clara County clearly functioned as a seedbed region for the young semiconductor industry, and the subsequent dispersal of production activities out of the valley has indeed coincided with the maturing of the industry. Vernon's observation that movement along the development cycle renders an industry increasingly footloose is, in this case, a valuable insight.

15. For a description of the evolution of the scientific-technological agglomeration around MIT and Harvard in the Boston-Cambridge area, which goes back to the post-WWI years, see Rand (1964) and Shimshoni (1967). The minicomputer industry, which boomed in the late 1970s, is the leading industrial sector in the Route 128 region. It has played the same role as the semiconductor industry in Silicon Valley. See Dorfman (1982) on the minicomputer industry.

16. Route 128 passes through 20 different towns and cities and through portions of 4 different counties.

REFERENCES

ADAMS, R. (1977) The Boston Money Tree. New York: Thomas Y. Cromwell.

BOUCHER, N. and D. DENISON (1982) "Tech towns." Boston Phoenix (January 12).

BYLINSKY, G. (1974) "California's great breeding ground for industry." Fortune 89: 128-135.

DORFMAN, N. (1982) Massachusetts High Technology Boom in Perspective. Cambridge, MA: MIT Center for Policy Alternatives.

Hilgenhurst Associates (1983) "Burlington Growth Impact Study: Final Report." Author.

HIRSCH, R. (1976) Social Limits to Growth. Cambridge, MA: Harvard University Press.

KELLER, J. (1979) Industrialization, Immigration, and Community Formation in San Jose, California: Social Processes in the Electronics Industry. Ann Arbor: University of Michigan.

Metropolitan Area Planning Council (1977) Residential Zoning in the MAPC Region. Boston: Author.

MUTLU, S. (1979) "International and interregional mobility of industrial capital: the case of the American automobile and electronics companies." Ph.D. dissertation, University of California, Berkeley.

RAND, C. (1964) Cambridge, U.S.A. Hub of a New World. New York: Oxford University Press.

ROBINSON, A. (1980) "Giant corporations from tiny chips grow." Science 208 (May 2): 480-484.

ROGERS, E. and J. LARSEN (1984) Silicon Valley Fever. New York: Basic Books.

Santa Clara County Housing Task Force (1977) Housing: A Call for Action. San Jose, CA: Santa Clara County Planning Department.

Santa Clara County Industry Housing Management Task Force (1979) Living Within Our Limits. San Jose, CA: Santa Clara County Planning Department.

Santa Clara County Planning Department (1979) Transportation/Land Use Planning Within the Present General Plans Structure. San Jose, CA: Author.

——(1978) 1975 Special Countywide Census. San Jose, CA: Author.

——(1980) Components of Yearly Population Increase, 1950-1979: Info Fact Sheet 660. San Jose, CA: Author.

——(1978) Total Population by Age and Sex: April 1950, 1960, 1970, 1975; Info Fact Sheet 591. San Jose, CA: Author.

——(1976) Assessed Valuation by City, Selected Years, 1959-1976: Info Fact Sheet 565. San Jose, CA: Author.

——(1972a) Socio-Economic Characteristics of Cities, April 1, 1970: Info Fact Sheet 469. San Jose, CA: Author.

——(1972b) Assesed Valuation for Selected Fiscal Years, 1959-1971, and Estimated Assessed Valuation Per Capita, 1959-1971, By City: Info Fact Sheet 389. San Jose, CA: Author.

SAXENIAN, A. (1980) "Silicon chips and spatial structure: the industrial basis of urbanization in Santa Clara County, California." Masters thesis, University of California, Berkeley.

SHIMSHONI, D. (1967) "Region development and science-based industry," in Meyer and Kain (eds.) Essays in Regional Economics. Cambridge, MA: Harvard University Press.

Stanford Environmental Law Society (1971) San Jose: Sprawling City. Palo Alto, CA: Stanford University.

U.S. Department of Commerce, Industry, and Trade Administration (1979) A Report on the U.S. Semiconductor Industry. Washington, DC: U.S. Government Printing Office.

VERNON, R. (1966) "International investment and international trade in the product cycle." Quarterly Journal of Economics 80: 190-207.

——(1960) Metropolis 1985. Cambridge, MA: Harvard University Press.

5

Defensive Cities:
Military Spending, High Technology,
and Human Settlements

ANN ROELL MARKUSEN
ROBIN BLOCH

☐ IN THE EARLY 1980s it became the fashion for state and local governments in the United States to create "high technology" promotional programs. High tech industries appeared to be the only life raft in a stormy sea created by rather persistent international economic crises, especially as the federal government showed no will to adopt a New Deal type of industrial policy or to assuage the United States's rapidly deteriorating balance of trade. State and local initiatives became predicated on the belief that the future viability of regional and urban economies lay with high technology sectors, conceived of as clean, innovative, entrepreneurial, and generating new technologies to improve human welfare. These "sunrise" industries, to borrow a phrase from Lester Thurow (1980), are often contrasted to the dirty, sluggish, bureaucratic, and conservative performance of management in the more mature, or "sunset," industrial sectors such as steel.

In parallel fashion, places that have posted a better-than-average rate of high tech job growth are often viewed as more dynamic and entrepreneurial than older, fiscally troubled industrial centers. Whereas geographers and economists once assumed that innovation would continue to emanate from older, cosmopolitan "seedbeds," recent scholarship has argued that the manufacturing belt is increasingly being challenged by newer metropolitan areas in the

West and South (Norton and Rees, 1979; Weinstein and Firestine, 1978; Hall, 1982). Causes of this new leadership have been variously ascribed to entrepreneurial preferences (Hall, 1982), the ossification of older areas under oligopolistic influences (Checkland, 1975; Markusen, 1984a), and the importance of specialized labor pools and amenities (Rees and Stafford, 1983; Joint Economic Committee, 1982).

However, one important source of high tech industrial stimulus seems to have been largely overlooked in these interpretations—the role of military spending.[1] In this chapter we argue that military spending in the United States has been a powerful industrial and regional policy that has profoundly affected the patterning of human settlements in the United States. Drawing high tech production toward the "defense perimeter," military procurement has spurred the industrially diverse manufacturing belt to create extensive, low-density, industrial park suburbs (Southwest Los Angeles, Silicon Valley, Dallas-Forth Worth's "Silicon Prairie," and Anaheim) adjacent to, but with relatively few political or cultural links to, older metropolitan areas. It has also created a newer generation of medium-sized detached metropolitan areas where, in a reversal of twentieth century urbanization tendencies, the local economy is highly dependent on one or a few sectors and demonstrates few tendencies toward diversification. The resulting geopolitical map is one in which highly military-dependent, relatively homogeneous and conservative cultural enclaves are counterposed to the strong industrial working class and bourgeois traditions of our large, mature manufacturing belt cities. We argue not only that the image of high tech production has been laundered (i.e., that its military roots and sustenance have been obscured), but that our most economically troubled communities have little prospect of capturing a share of this type of economic growth.

HIGH TECH AS MILITARY PRODUCTION

The phrase "high tech," as an adjective and even as a noun applied to individual industries, has been used quite loosely by academics, the popular press, and policy makers alike. It generally connotes one or more of three features: an extensive degree of technical sophistication embodied in a product, a rapid rate of employment growth associated with an innovative product, and a large research and development effort associated with production.[2] However, not all sectors producing a sophisticated product or engaging in extensive R&D are growing rapidly, and vice versa. In a related research effort we have created an operational definition encompassing both the product sophistication and the research-intensive features, which classifies as "high tech" those industries with a higher-than-average proportion of their workforce in scientific and technical occupations (engineers, engineering technicians, computer scientists, life scientists, mathematicians). This rather generous definition produces a set of 100 4-digit manufacturing sectors.

Whether we conceive of high tech as a sophisticated product, a substantial R&D commitment, or a highly engineering-oriented labor force, we would expect military equipment, weapons, and transport to be highly high tech in nature. Cold war rivalries have resulted in continual innovation in weaponry and delivery systems. The goal of military R&D is to make the product of the opposition ineffective, and the result of international competition of this sort, particularly between the Soviet Union and the United States, has been the creation of a dynamic whereby existing products are rapidly rendered obsolete. Product innovation, then, rather than process innovation, dominates industrial research and development activities in these sectors, creating what Malecki (1983: 91) calls technology push in place of demand pull as the growth trajectory. These sectors are thus more or less permanently locked into the initial stages of their product cycles, unless an entire system is displaced by another (such as battleships by submarines and aircraft).

Because of this constant product innovation and because the products themselves are highly sophisticated, the labor force is distinctively skewed toward highly skilled workers. Engineers alone accounted for relatively larger percentages of the workforce in aircraft (11%), communications equipment (15%), guided missiles (31%), and ordnance (5%) than in manufacturing as a whole (3%) (De Grasse, 1983: 33). One researcher found that 59% of all aeronautical engineers, 38% of physicists, 22% of electrical engineers, 20% of technical engineers, 20% of mechanical engineers, and 20% of metallurgical engineers were employed as civilians either directly by the Defense Department or by defense-related industries.[3] Given the heavy reliance of military producers on this type of scientific personnel, these sectors are apt to figure prominently among high tech industries, however defined.

The contemporary success of high technology industry is intimately connected to defense spending, even by the Reagan administration's accounts. The Bureau of Industrial Economics

TABLE 5.1
Defense-Related High Tech Industries

SIC#	Industry Name	Employment 1977	Defense Output Share 1982 (%)	Defense Output Share 1987 (%)	Defense Output Growth 1982-87 (%)
2812	Alkalies and chlorine	11,831	5.6	7.1	53.3
2813	Industrial gases	7,332			
2816	Inorganic pigments	12,003			
2819	Inorganic chem. misc.	78,192			
2865	Cyclic crudes, dyes	35,499			
2866	Organic chem. misc.	11,242			
2892	Explosives	11,546	34.3	41.2	58.7
3482	Small arms ammunition	12,199	25.0	39.9	129.5
3483	Ammunition, ex small arms	20,589	90.9	93.2	55.6
3484	Small arms	17,495	13.8	6.5	−43.3
3489	Ordinance	19,042	79.7	81.2	35.3
3541	Machine tools, cutting	59,432	6.2	7.5	54.4
3542	Machine tools, forming	23,154	4.8	6.3	70.0
3544	Special dyes, jogs, molds	106,175	6.0	7.5	45.4
3545	Machine tool accessories	54,177			
3562	Ball, roller bearings	50,286	5.8	6.8	45.4
3573	Electronic computing equip.	192,510	7.1	12.7	141.0
3624	Carbon, graphite products	12,086	7.7	9.3	51.4
3622	Radio, TV communications	333,006	58.0	62.5	54.2
3671	Cathode ray tubes	36,808	7.3	11.5	105.3
3674	Semiconductors	114,011	12.5	12.5	51.4
3675	Electronic capacitors	28,647	17.0	19.8	49.3
3676	Resistors, electronic	24,918			
3677	Resistors, electric	22,424			
3678	Connectors	26,020			
3679	Electronic components	125,998			
3721	Aircraft	222,805	40.4	46.1	58.7
3724	Aircraft engines	106,322	53.5	56.1	32.9
3764	Missile, space engines	17,011			
3728	Aircraft parts	101,900	41.2	44.2	34.9
3769	Missile, spacecraft parts	10,189			
3761	Missiles, space vehicles	93,933	67.5	79.4	64.4
3795	Tanks, tank parts	12,122	93.8	95.0	47.2
3811	Engineering instruments	42,178	27.7	33.6	59.9
3825	Instruments measuring elec.	66,622	8.4	9.8	49.7
3832	Optical instruments	29,883	28.0	30.7	38.0

SOURCES: Census of Manufactures (1977) as tabulated in Glasmeier et al. (1983, Table 1 :5) and Henry (1983, Table 3: XLVII). The BIE estimates reported by Henry are sometimes aggregated into three digit SICs, thus the shared percentages in the later three columns.

concluded in 1983 that this country's defense industrial base consists of 58 sectors, almost all of which are manufacturing activities.[4] Almost all fall into the high tech category, shown in Table 5.1 with their recent employment levels, defense market shares, and expected defense-related output growth rates.

It is fairly easy to demonstrate that heavily defense-dependent manufacturing sectors are dominated by those in the high tech category and that a substantial portion of high tech industries are important military suppliers. First, the top 20 of BIE's defense-dependent manufacturing sectors include only 2 sectors not classified as high tech: shipbuilding and repair and miscellaneous nonferrous foundries. Second, while the 36 sectors in Table 1 encompassed less than one-half of the 100 high tech manufacturing sectors, they accounted for 47% of all high tech manufacturing employment in 1977, or approximately 2.25 million jobs.

By 1984 these military-based high tech sectors will undoubtedly account for an even higher percentage of all high tech industries than they did in 1977. The extraordinary impact of the Reagan buildup is shown by the defense output growth rates predicted by the BIE, shown in the final column of Table 1. The military-purchased output of most sectors will grow by more than 50% in the 5-year period. As a result, the defense share of output will rise dramatically. The biggest share shifts will occur in ammunition, missiles, aircraft, explosives, engineering instruments, electronic tubes, computers, and radio and TV equipment. By 1987 these high tech sectors will be significantly more defense dependent than they were in either 1979 or 1982.[5]

THE LOCATIONAL TENDENCIES
OF THE MILITARY-INDUSTRIAL COMPLEX

Until the latter half of the nineteenth century, military conflict was a strong reinforcer of urbanization patterns. Because war was a labor-intensive activity, and because weapons were generally built to human scale (armor, pikes, muskets), cities from the medieval period onward sported fortifications, garrisons, armories, and quarters for the manufacture of guns, ammunition, and other carefully crafted equipment. Only with the shift to capital-intensive techniques of warfare, especially the development of the warship and the airplane, did production of large-scale weapons systems begin to shun cities for more peripheral locations. This antiurban centrifugal dispersion appeared simultaneously with the shift from state to private sector production, embodied in the "defense contract."[6]

Why should military-oriented manufacturing exhibit spatial tendencies so unlike production geared to nonwar markets? The answer is that these sectors (such as military construction, aircraft, ammunition, ordnance, tanks, ship building, missiles, communications equipment, electronics, explosives, and scientific instruments) operate in markets that differ markedly from sectors competing in commercial markets. First of all, the nature of demand differs dramatically from other manufacturing sectors. Military goods are sold to one monopsonist buyer, the federal government. But, unlike private sector monopsonists, this customer does not use its market power to depress the price of the product (and therefore the profits of sellers) in the market transaction; demand is highly insensitive to price changes. Numerous authors have stressed, instead, performance and timeliness as dominant selection criteria for the Department of Defense awards. Cost insensitivity has been institutionalized into the cost-plus contract, in which overruns in excess of 100% are common in larger weapons and equipment procurement programs (Baldwin, 1967: 80-89; Gansler, 1980: 89-92; Melman, 1974: 27-39).

On the supply side, these industries are highly oligopolistic. A small number of corporations do the bulk of Defense Department business (Adams, 1981: 33-41). The top 100 companies account for 70% of total business, the top 25 for 50%, and the top 5 for 20% (Gansler, 1980: 36-45). Concentration ratios at the sectoral level are high and have been increasing over time (Baldwin, 1967: 62-78). At the level of individual products, firms generally compete with only one or two others for the initial contract for a new system and, after that, exercise a virtual monopoly. Subcontracting is considerably more competitive, but in this arena, too, concentration has been growing rapidly, in part because the larger prime contractors have increasingly chosen to produce in-house rather than to buy from others. Often subcontractors have survived by occupying market niches and becoming monopolists of certain highly specific parts (Gansler, 1980: 128-44).

The nature of the military product and its production process is also quite distinct from commercial manufacturing. The units of output of assembled weapons, equipment, and transport systems are often (and increasingly) few in number and highly sophisticated in nature. Constant innovation requires numerous platoons of scientific and technical personnel engaged in design, prototype building, and testing and skilled labor for assembly. Often immense in size (e.g., submarines, military transport planes, missiles), these products tend to be produced in very large-scale facilities. Batch, rather than assembly-line, production predominates, so that defense-oriented plants are more like huge craft shops than standardized mass production factories.

Resource and market arrays, important in an earlier era and especially for industries such as steel, are relatively unimportant locational factors for military-oriented producers. Materials account for a small share of total costs, and those used tend to be high in value relative to transport costs. Markets are ephemeral. In one sense, the Department of Defense in Washington is the market, but military products are not delivered there. Some assembled products, such as ships, submarines, and aircraft, literally transport themselves to market. Most other military products are stockpiled and stored for some future mobilization, but the location of these receptacles may be dispersed, secret, or variable. Missile silos are a good example. As most military material and equipment is indeed meant to be mobile, the physical link to markets is not very important, at least for the larger products such as tanks, missiles, and aircraft.

Two other inputs rank high on the list of locational factors: land and labor. Land is important because the immense size of some plants requires extensive space for construction. Furthermore, the innovative nature of many products requires significant on-site experimentation and testing, so that additional space must be available for this purpose. In the case of large, mobile products (such as fighters, bombers, and transport planes) this space need is extraordinary. For highly secretive or potentially lethal products, experimentation requires extensive, unpopulated land areas. Cheap, buildable land may also be desired for construction of low-cost housing for a resident workforce. An enterprising management may also be interested in profiting from the development of land around the new site. Thus, military-oriented entrepreneurs may favor areas with large tracts of undeveloped land amenable to both industrial and residential use. The attractiveness of Los Angeles to the aircraft industry, which was relocating from the East in the 1930s, may be due in large part to these land-related factors.[7]

Labor is also an important locational requirement, as payroll forms a large share of total costs. Qualified scientific and skilled labor ensure companies a competitive edge in meeting performance demands of the client. Yet, for a number of reasons, labor may be more mobile than other factors in the special case of military-oriented sectors. Because cost competition is not a major aspect of market conduct, and because a standard rate of return may be charged on all legitimate costs, military contractors may prefer to recruit labor from other regions, paying for relocation costs. Large pools of engineers and scientists have been assembled in this way in places like Silicon Valley and Los Angeles. In war time, skilled blue-collar labor has been similarly recruited, especially to shipyards in Vallejo, Richmond, and other West Coast ports.

Recruiting and building a new local labor supply offers certain advantages to military-based industries. It may discourage unionization. Defense contractors may not care terribly about wage costs, but work stoppages, union-won work rules, and efforts to improve working conditions may hamper the ability of a company to complete a contract in a timely fashion. The desire to create a new workforce far from the centers of older industrial society may also be an expression of the conflict between military attitudes toward manpower and civilian, corporate ones. Collective bargaining, responsiveness to financial incentives, and labor mobility are aspects of capitalist labor relations inimical to the military mode, which has a completely internal labor market with advancement based on merit and an emphasis on loyalty to the organization above any individual interest (Fallows, 1981: 107-114). The frequent movement of management between the Defense Department, the armed services (retired personnel), and military-oriented firms may result in the carrying over of these values into the location decisions of the latter. This cultural factor may also explain why military-oriented industries may be willing to dominate a local labor market in a medium-sized or even small town, in much the same way that the military brass dominate self-contained bases.

Finally, agglomeration economies are a powerful factor in the location of military-oriented manufacturing. They give the military-industrial complex its spatial analog. These agglomerations may be of two types. First, defense-oriented production locates in close proximity to existing military bases, for several reasons. For instance, the experimental nature of the product may require collaboration between base personnel and producers (test pilots, for example). Second, military bases may provide a regular stream of trained, specialized workers mustered out after short- (2 to 3 years) or medium-term (20 years) service. As these discharged soldiers generally are not paid to return to their hometowns (although they are paid for their trips to bases to begin with), significant numbers of them choose to remain in the vicinity (Long, 1976).[8] Third, military bases sometimes provide free infrastructure to firms producing nearby.

Government facilities of other types may also serve as magnets for military-oriented production. The location of government R&D labs is significantly more skewed toward the South and the West than is private corporate R&D (Malecki, 1981). Much of this R&D is military-oriented, and commercial spinoffs or suppliers may locate nearby.

The second set of intersectoral connections consists of the pull exerted by certain pioneer sectors on their suppliers. Decisions to locate the production of large weapons and transport systems outside of major existing industrial complexes has stimulated sequential shifts by related sectors to those sites, both to be near their performance-oriented clients and to draw upon the new skilled labor that specializes in the defense industry's needs. An example of such a sequence is the creation of new machining, steel, metal fabrication, communication, and missiles sectors in Southern California following the 1930s' decisions by airframe manufacturers to relocate from New York and other Eastern cities.

It is important to note that the federal government has often played a direct role in the location of productive capacity in these sectors. Many defense-oriented manufacturing activities take place in government-built and owned facilities. Wartime imperatives explain a fair amount of this investment, but some plants have been built in peacetime as well. While many government-built plants have been sold to private firms, other remain government-owned but contractor-operated (GOCOs). Still others continue to be operated as well as owned by the government itself. The location decisions of the federal government, both in siting bases and productive manufacturing capacity, are often affected by strategic considerations not raised in this discussion. An example might be the opening of the Pacific front in World War II, with its dramatic impact on Pacific coast military staging and manufacturing. However, many of the considerations suggested to affect private firms' location decisions may also operate within the Department of Defense.

In summary, buildable land and the proximity of military bases are the chief locational factors explaining the long-term orientation of American military-oriented manufacturing. These factors have pulled the major heavy military systems plants toward sunbelt locations and, to a lesser extent, to peripheral sites within the manufacturing belt. In turn, a manufacturing labor force has been drawn and anchored around these sectors and a number of supplier industries, some old and some new, have attached themselves to these centers.

The large, spatially concentrated complexes that have resulted from these tendencies have fundamentally altered the human settlement pattern within the United States. Located outside the older manufacturing cities, due to the attraction of large parcels of undeveloped land, to the proximity of existing military bases, and to the repulsion of an organized, militant industrial workforce, they have spawned new peripheral communities that employ disproportionate numbers of engineers and certain strata of skilled workers, predominantly male. These communities are marked by relatively conservative, patriarchal, and promilitary cultural attitudes. They may be adjacent to older metropolitan centers, but form few interconnections with them, coexisting as distinct labor markets and cultural enclaves. In essence, they are industrial "cities" of an entirely new type, anticosmopolitan, antilabor, antiliberal.

DEFENSE ENCLAVES: SOME EXAMPLES

Since World War II new territorial complexes have risen on the defense perimeter, notably in the Southwest. The growth of these defensive cities has been crucially (although not solely) dependent on military procurement and production, and their political culture reflects these defense origins. Despite the relative lack of systematic attention given to the role of post-World War II military expenditure in shaping the pattern of human settlement in the United States, and the social and economic consequences of this, examples of such defense enclaves are not difficult to isolate and describe. Here, we will briefly provide such examples, concentrating on the three most populous sunbelt states: California, Texas, and Florida.

California's postwar economic growth has, to a large extent, been based on the defense industry. The creation of Southern California's aircraft industry in the 1920s and 1930s laid the foundation for the War Production Board-financed and organized military aircraft industry in the next decade. This, in turn, acted as a basis for the cold war-fueled expansion of the aircraft industry into aerospace (missiles being crucial after the mid-1950s) in Southern California. The aerospace industry effectively gave birth to new settlements that acted as magnets for in-migration. California showed three times the population growth of any other state in the 1950-1960 period, and net migration into California was strongly correlated with cycles in defense contracting in the postwar period (Clayton, 1962: 289-290; Dietrick, 1984). This growth occurred in Los Angeles County initially, but fanned outward as land became scarce. Southwest Los Angeles—Hawthorne, El Segundo, Culver City—thus became a center for aerospace production as did sections of the San Fernando Valley (Burbank and Van Nuys, and later Canoga Park and Woodland Hills), the West Covina and Azusa area, Pomona and Ontario, and, further out into the Mojave, Palmdale (near Edwards Airforce Base). The spillover into Orange County began in the mid-1950s, with Anaheim and Fullerton benefitting particularly. By the 1970s and 1980s the Los Angeles region housed almost all of the large aerospace manufacturers—Hughes Aircraft, Lockheed, Rockwell International, Northrop, Mc Donnell Douglas, TRW, and General Dynamics—and many subcontractors (Scheer, 1983: 12-14, Clayton, 1962: 290).

Defense production also shaped the postwar development of other cities in California. The aircraft industry in San Diego was supplemented with missile production and communications. Portions of Santa Clara County—later Silicon Valley—such as Sunnyvale, home of Lockheed's missile division, and Mountain View also gained dramatically, as did the nuclear weapons research and development center of Livermore (Markusen, 1984b; Clayton, 1962).

A similar tale can be told for much postwar urban and regional development in Texas and Florida. The lobbying of a congressional delegation led by Sam Rayburn and Lyndon Johnson resulted in the opening of a B-24 plant in Forth Worth by General Dynamics in 1942. Over the next decades General Dynamics was joined by such aerospace and defense electronics companies (the latter particularly from the early 1960s onward) as Lockheed, Rockwell, LTV (now the Vought Aerospace division), Bell Helicopter, Texas Instruments, and Electrospace Systems and their subcontractors and spinoffs. New defense-based communities such as Arlington, Richardson, Irving, and Grand Prairie sprang up on the wide open spaces of what has come to be called the Dallas-Fort Worth Metroplex, or "Silicon Prairie." Defense companies also located in San Antonio, the site of more military bases than any other city in the United States, and, more recently, in Austin (Hurt, 1984: 135, 238-240; Butts, 1984: 9-11). Equally, the growth of Florida's Golden Girdle (the area between Tampa-St. Petersburg on the Gulf Coast and Daytona Beach to the Palm Beaches on the Atlantic) and "Silicon Coast" (West Palm Beach through Miami) has also been shaped by the imperatives of defense production. The communities of Titusville, Cocoa Beach, and Melbourne, situated around the Cape Canaveral-Kennedy Center-Patrick Airforce Base space complex, as well as Orlando, St. Petersburg, Tampa, and West Palm Beach, have come to house such defense contractors as Martin Marietta Aerospace, Harris Corporation, United Technologies' Pratt & Whitney and United Space Boosters, Honeywell, Rockwell International, Grumman, Mc Donnell Douglas, and Racal-Milgo. Growth in the last few years, particularly in defense electronics and communications, has been tremendous (Aviation Week and Space Technology, 1983: 63).

These defense-based settlements in California, Texas, and Florida are not the only examples of the new military cities of the postwar period. Running right along the defense perimeter from New England through the Southeast, the Southwest, and the Pacific and Mountain states are other military-based dependent high tech enclaves—Hartford, Connecticut; Marietta, Georgia; Huntsville, Alabama; Oak Ridge/Knoxville, Tennessee; Phoenix; Denver; and so on. After the hard times of the early and mid-1970s—with defense cutbacks causing often massive dislocation in many of the communities mentioned above—it is now a boom period for defensive cities, which now form the spatial expression of a Reagan administration "industrial policy" centered around military spending (Markusen, 1984c). Indeed, such development seems likely to continue and expand in space through the 1980s—with aerospace companies such as Grumman, Sikorsky, and Beech increasingly moving production facilities to or building new plants in the Southeast, particularly in Alabama and Tennessee, while a similar shift occurs across the country from the Los Angeles region into the Mojave and Arizona. Such development will continue to exclude Northeast and Midwest locations—areas that, according to a recent editorial in an aerospace trade journal, are typified by "an anti-military climate . . . on top of a high local tax situation," and thus have no locational attraction for defense contractors (Gregory, 1984: 1.).

In pointing to the hitherto relatively ignored significance of defense expenditures for the creation of high tech industry and the spatial consequences of this, we are aware that far more work, simultaneously conceptual and empirical, needs to be done. The growth of defensive cities must be located within the body of existing accounts of postwar suburbanization and the frostbelt/sunbelt shift. Equally, work must be continued on the economic and cultural charateristics of defense enclaves. Certainly, it would initially appear that women and minorities have not (except during the war years) and are not sharing in the high end of the defense employment boom. A recent study at the University of California, Berkeley, on the effects of military spending on urban development in California points, for example, to an underrepresentation of Hispanics in aerospace employment in the Los Angeles region and lower wages (61% of those of men) for the women who form only 26% of the aerospace workers in Los Angeles (Markusen, 1984b). Moreover, the potential of new materials and automated manufacturing processes for repelling blue-collar metal-working employment in the aerospace industry is not inconsiderable in the medium term. The cultural distinctiveness of the defense enclaves also deserves much attention. To the high-skill/low-skill occupational and cultural divisions widely perceived as characteristic of high tech and defense-driven industries can be added another division: the spatial and cultural bifurcation of middle-class culture, with professionals located in the "cosmopolitan" inner cities (or, more pertinently here, within the inner ring freeways of sunbelt metropolises) and the scientific, technical workforce located on the fringes in defense enclaves.

NOTES

1. For instance, in the Rees and Stafford (1983) review of the contribution of regional growth and industrial location theory toward understanding the development of high technology complexes, there is no mention of defense spending. For exceptions, see Stein (1983) and Malecki (1983).

2. Glasmeier et al. (1983) discuss these connotations and the difficulties of operationalizing each of them.

3. In skilled blue-collar categories, military-related production accounted for 54% of tool and die makers, 25% of pattern and model workers, 25% of sheet metal workers, 19% of machinists, and 19% of airplane mechanics, and 14% of draftsmen (Rutzick, 1970: Table 2).

4. The exceptions are several mining, construction, and manufacturing sectors. The BIE defense shares represent direct and indirect shares of industrial output estimated from updated and adjusted coefficients from the 1972 national input-output model. The BIE method includes only Department of Defense outlays and, thus, significantly underestimates military demand; it does not capture the nuclear activities of the Department of Energy, the defense aspects of National Aeronautics and Space Administration, expenditures of the National Guard, Coast Guard, Selective Service, Federal Emergency Management Agency, nor the stockpiling activities of the General Services Administration (Shutt, 1984; Kennedy, 1983: 50-51).

5. In the case in industries such as electronics and computers, this reverses a postwar trend in which commercial applications have grown faster than military ones (Saxenian, 1980).

6. For an elaboration of this historical progression, see Markusen (1984a).

7. The aircraft industry located mainly on the periphery of Los Angeles. By 1939 only 7% of all aircraft workers were employed in cities of more than 500,000, compared to 23% of all industrial workers; 62% of all aircraft workers were employed in communities of less than 25,000, compared to 40% of all manufacturing. Numerous bedroom communities sprang up around the new aircraft fringe of Los Angeles (see Cunningham, 1951: 58-65).

8. Florida's Education Department, for example, has a computer program listing the skills and availability of 13,000 military retirees in the state for the use of high tech recruiters (Aviation Week and Space Technology, 1983: 63).

REFERENCES

ADAMS, G. (1981) The Iron Triangle: The Politics of Defense Contracting. New York: Council on Economic Priorities.

Aviation Week and Space Technology (1983) "Florida's business climate attracting new industry." (February 21).

BALDWIN, W. L. (1967) The Structure of the Defense Market, 1955-1964. Durham, NC: Duke University Press.

BUTTS, N. (1984) "Arms of war: made in Texas by Texans." Texas Observer (February 24).

CHECKLAND, (1975) Under the Upas Tree. Glasgow: Glasgow University Press.

CHINITZ, B. (1960) "Contrasts in agglomeration: New York and Pittsburgh." American Economic Review, Papers and Proceedings: 279-289.

CLAYTON, J. L. (1962) "Defense spending: key to California's growth." Western Political Quarterly 15(2).

CUNNINGHAM, W. G. (1951) The Aircraft Industry: A Study in Industrial Location. Los Angeles: Morrison.

DEGRASSE, R., Jr. (1983) Military Expansion, Economic Decline: The Impact of Military Spending on U.S. Economic Performance. New York: M. E. Sharpe.

DIETRICK, S. (1984) "Military spending and migration into California." Berkeley: University of California, Department of City and Regional Planning. (unpublished)

FALLOWS, J. (1981) National Defense. New York: Random House.

GANSLER, J. (1980) The Defense Industry. Cambridge, MA: MIT Press.

GLASMEIER, A., P. HALL, and A. MARKUSEN (1983) Defining High Technology Industries: Working Paper 407. Berkeley: University of California, Institute of Urban and Regional Development.

GREGORY, W. (1984) "Recruiting high technology." Aviation Week and Space Technology (February 27).

HALL, P. (1982) "Innovation: key to regional growth." Transaction/Society 19(5).

HURT, H. (1984) "Birth of a new frontier." Texas Monthly (April).

Joint Economic Committee (1982) Location of High Technology Firms and Regional Economic Development. Washington, DC: U.S. Government Printing Office.

KENNEDY, G. (1983) Defense Economics. London: Duckworth.

LONG, J. (1976) "Interstate migration of the armed forces." Presented at the Annual Meeting of the Southern Sociological Society, Miami.

MALECKI, E. (1983) Federal and Industrial R&D: Locational Structures, Economic Effects, and Interrelationships; Final Report. Washington, DC: National Science Foundation, Division of Policy Research and Analysis.

——— (1981) "Government-funded R&D: some regional economic implications." Professional Geographer 33(1): 72-82.

MARKUSEN, A. (1984a) Defense Spending and the Geography of High Tech Industries: Working Paper 423. Berkeley: University of California, Institute of Urban and Regional Development.

——— (1984b) Military Spending and Urban Development in California: Introduction to Study Report. Berkeley: University of California, Department of City and Regional Planning.

——— (1984c) Defense Spending: A Successful Industrial Policy? Working Paper 424. Berkeley: University of California, Institute of Urban and Regional Development.

MELMAN, S. (1974) The Permanent War Economy: American Capitalism in Decline. New York: Simon & Schuster.

NORTON, R. and J. REES (1979) "The product cycle and the spatial decentralization of American manufacturing." Regional Studies 13: 141-151.

REES, J. and H. STAFFORD (1983) A Review of Regional Growth and Industrial Location Theory: Towards Understanding the Development of High-Technology Complexes in the United States. Washington, DC: Office of Technology Assessment.

RUTZICK, M. (1970) "Skills and location of defense-related workers." Monthly Labor Review 93: 11-16.

SAXENIAN, A. (1980) "Silicon chips and spatial structure: the semiconductor industry and urbanization in Santa Clara County, California." Master's thesis, University of California, Berkeley.

SCHEER, R. (1983) "California wedded to military economy but bliss is shaky," in Servants or Masters? Revisiting the Military Industry Complex (Special Supplement). Los Angeles Times (July 10).

SHUTT, R. (1984) The Military in Your Backyard: A Workbook for Determining the Impact of Military Spending in Your Community. Mountain View, CA: Mid-Peninsula Conversion Project.

STEIN, J. (1983) "Planning in the nuclear age: waiting for apocalypse." Atlanta: Georgia Institute of Technology, Graduate City Planning Program. (unpublished)

THUROW, L. (1980) The Zero-Sum Society. New York: Basic Books.

WEINSTEIN, B. and R. FIRESTINE (1978) Regional Growth and Decline in the United States: The Rise of the Sunbelt and the Decline of the Northeast. New York: Praeger.

Part III

The Transformation of Services

6

The Changing Fortunes
of Metropolitan Economies

THOMAS M. STANBACK, Jr.

□ AFTER MORE THAN a quarter-century of rapid advances in computer and related technology, there is abundant evidence that the pace of change, far from slowing down, is accelerating, giving promise of a broadening scope of applications in the private and public sectors and in the home. In what way is the role of metropolitan economies within the larger national economy being changed by these developments, and what new problems and opportunities are being created?

In addressing these questions, three propositions are set forth: (1) that the impact of the new technology cannot be understood without recognizing the nature of the forces that have brought about the rise of services, an economic transformation in which technology has played an important but only contributory role, and that has, itself, altered the role of metropolitan economies: (2) that although adoption of computer-related technology throughout the American economy today is uneven, there is evidence that we have entered a new phase of rapidly accelerating and increasingly sophisticated applications that will involve a broad spectrum of users; and (3) that the impact of the new technology involves changes in the nature of work and the organization of user firms with important implications for the location of economic activity and, accordingly, for metropolitan economic development.

THE RISE OF SERVICES[1]

During the postwar years services have grown from 57% of employment to more than 70% today, accounting for virtually all job increases since the beginning of the 1970s.[2] This rapid transformation has not involved all services equally, however. The expansion of service employment has been accounted for chiefly by the nonprofit services (education and health), government, and, not widely appreciated, the producer services (finance, insurance, real estate, and other business services such as law, accounting, advertising). Closely associated with the latter, but not revealed directly by the employment data, has been a rapid growth of producer servicelike activities performed *within* corporate organizations of goods and service firms alike.

Similar rates of employment growth have not been experienced in the distributive services (wholesaling, transportation, communications, utilities), although there has been no decline in their relative importance when measured in value-added terms. Retailing has grown more or less in step with the overall economy, and consumer services have declined somewhat, largely because of reductions in private household services (there have been sharp increases elsewhere within the sector). The goods producing sector (agriculture, mining, manufacturing) has, of course, declined in terms of employment share; but it has not declined in value-added terms, reflecting major advances in productivity.

FACTORS CONTRIBUTING TO THE RISE OF SERVICES

The rise of the services has involved a major transformation in both *what* is produced and *how* it is produced. In terms of final products (what we produce) there has been a marked trend toward greater "product differentiation" and proliferation of new products and services, reflecting higher levels of income and changing tastes. But the growth in government and nonprofit services has been the principal source of change in the final demand for services. In spite of some retrenchment in the public sector in recent years, it is clear that the affluent, complex modern society requires major flows of heatlh, education, regulatory, and other services.

The transformation in how we produce has involved, on one hand, changes of production in manufacturing and, on the other, a rising importance of producer services (both free-standing and in-house). Changes in manufacturing production have been made possible by improvements in factory technology, transportation, communication, and management science and have brought a reduction in blue-collar production employment and a continuous shifting of plants—away from cities and (more or less in stages) to suburbs, smaller urban places, rural areas, and overseas. At the same time there has been a shift in the importance of functions other than direct production, involving greater emphasis on finance, marketing, product development, personnel management, and problems related to coping with a more complex regulatory environment. It is this shift in functional emphasis that accounts for the rise in producer services and for the shift of resources toward headquarters, sales, and other servicelike activities within the firm.

A number of forces have driven the changes sketched above, among which the rise of computer and related technologies is but one. Increasing size of markets, made possible not only by a growing population and per capita income but by a breaking down of regional barriers and new trends toward internationalization, has opened up new demands for product differentiation, branding, and sales promotion and has brought about a host of new products. Improved transportation and communications have facilitated the delivery of products to these broad markets through more elaborate but also more efficient distribution systems, and have made it economical to locate manufacturing plants in nonmetropolitan areas.

The rise of the large corporation requires special comment as a major factor in the service transformation. There has been so much said in recent years regarding the importance of small businesses in generating job increases that the role of large corporations has tended to be downplayed. Large companies have not changed substantially in importance in value-added terms since the 1960s-(Harris, 1983). They hold center stage in any analysis of the impact of the rise of services and changing technology on metropolitan economies. It is the large corporation that has played the principal role in reaching out to national and international markets as well as in providing major shares of banking, insurance, transportation, and communications services. Moreover, in terms of the economies of metropolitan areas, large corporations play a strategic role. The siting of their activities influences the export base of the host city both directly through adding to employment and income and indirectly through attracting support services (including, especially, producer services) and by enlarging demands for residentiary activities.

Finally, computer and telecommunications technologies have played an important role in the services transformation. This has been, in large measure, because application of these technologies has made it possible for large corporations to take on their expanded role by enabling them to process the vast quantities of data involved and to rebuild corporate structures in such a way as to organize and administer the new production, marketing, and developmental efforts. But technology has also made it possible for certain medium-sized and even small producer service firms (e.g., consulting, engineering, advertising) to provide for the growing needs for expertise and specialized servicing among both private sector firms and governmental agencies.

THE CHANGING ROLE
OF METROPOLITAN ECONOMIES

One of the major contributions of urban geographers and economists has been to make explicit the role of the export base in the economies of cities and towns. Urban places must export goods or services in order to import in turn, and labor, management, and other factors of production must be engaged in these activities as well as in local sector (residentiary) activities that service the local populace.

Cities have traditionally played roles as centers for providing distributive, governmental, and nonprofit services to their hinterlands, but the relative importance of these export roles has varied widely among places. During the era of industrialization many cities, both large and small, became manufacturing centers with major portions of their workforces devoted to production or related activities, while other cities continued to play roles as distribution centers for regional agricultural economies, as government centers, and so on.

As services have increased in importance, and as goods production has been relocated, many cities previously specializing in manufacturing have faced major problems of adjustment, and all cities have faced major changes in their industrial structures as certain types of service activities have taken on increased importance.

In short, the growth of services has brought about a new competition among metropolitan economies in which some have clearly been favored over others. Evidence of such competitiveness can be gleaned from an earlier study, updated by use of more recent data. In this study (Noyelle and Stanback, 1984) the 140 largest SMSAs were classified according to the industrial composition of employment and key characteristics of specialization in terms of business services, corporate headquarters, distribution, communications, and transportation. The principal groups of SMSAs are as follows:

(1) Nodal
  (a) National nodal (New York, Los Angeles, Chicago, San Francisco)
  (b) Regional nodal (e.g., Philadelphia, Houston, Atlanta)
  (c) Subregional nodal (e.g., Memphis, Syracuse, Charlotte)
(2) Functional Nodal (e.g., Detroit, Hartford, Rochester)
(3) Government-education (e.g., Washington, D.C., Albany, Harrisburg)
(4) Education-manufacturing (e.g., New Haven, South Bend, Ann Arbor)
(5) Production
  (a) Manufacturing (e.g., Buffalo, Flint, Greenville)
  (b) Mining-industrial (e.g., Tucson, Duluth, Charleston, West Virginia)
  (c) Industrial-military (e.g., San Diego, Norfolk, Huntsville)
(6) Residential-resort (e.g., Nassau-Suffolk, Tampa, Las Vegas)

Nodal places are service centers in which exports are concentrated primarily in distributive and producer services and, often, secondarily, in other services as well (e.g., nonprofit services, arts, or recreation). Many, but by no means all, are state capitals and, thus, are also heavily engaged in delivering public sector services. Usually, the degree of diversification and specialization among nodal centers is a function of their population size and the size of the markets they serve. The presence of corporate headquarters or other administrative installations is of considerable importance in explaining the structure of their economy. The classification of these nodal centers under three headings—national, regional, and subregional—captures much of the variations that result from market size differences.

Functional nodal centers are places specializing in both manufacturing production and selected service functions of the large corporation, mostly R&D and administration of large industrial divisions. Although they can be quite large—many are comparable in size to the regional and subregional centers—they are much more restricted in their service functions than the nodal centers.

Government-education places are, for the most part, state capitals, seats of large educational institutions, or both.

Education-manufacturing places are predominantly old industrial centers that are also sites of university complexes.

TABLE 6.1

	1969-1976 [a]	1976-1983 [b]
National Nodal	9	6
Regional Nodal	4	3
Subregional Nodal	6	5
Functional Nodal	7	9
Government Education	3	4
Education-manufacturing	5	7
Manufacturing	10	10
Industrial-military	2	2
Mining-industrial	8	8
Resort-residential	1	1

a. Based on data analyzed in Noyelle and Stanback (1984).
b. Based on Table 1.

Production centers include manufacturing, mining, and industrial-military places. They are characterized by more routine production than that found in functional nodal places and are singularly weak in attracting administrative or research establishments of industrial corporations. Industrial-military places differ significantly from other production centers in that they are characterized by a large presence of federal government, civilian, and military employees on military bases, shipyards, and so on.

Finally, *resort-residential* centers include some of the metropolises that outlie the large national nodal centers, as well as many places that have developed since World War II as resort and retirement centers.

Comparison of typical growth experience of each group sheds considerable light on what has been taking place. The rankings of the several groups of SMSAs according to average net employment growth are shown in Table 6.1 for 1976-1983 (based on Table 6.2) and for the preceding period, 1969-1976.

An initial observation is that, in terms of rates of change in total employment, the experience of the several groups has shown considerable variation. By far, the most rapid growth has been experienced by places that have been favored by the new postwar patterns of agglomeration: people-centered activities and the rise of new light industries (resort-residential), military build-up (industrial-military), activities related to corporate offices and business, financial or distributive services (regional and subregional centers), or activities strongly influenced by higher government or educational-centered development (government-education places). The slowest growing places are those that are still most clearly identified with the earlier industrial era, manufacturing places, functional nodal, and mining-industrial places.

TABLE 6.2
Average Annual Rates of Employment Change:
Total Employment and Selected Industry Groups,
by Type of Metropolitan Economy, 134 Larges SMSAs
1976-1983

Type of SMSA*	Total	Whole-Mfg.	Whole-retail	FIRE	Other Services**	Gov't
National nodal (4)	1.37	−.92	1.45	3.36	4.13	−.16
Regional nodal (19)	2.64	.73	2.84	3.61	4.85	1.24
Subregional nodal (16)	2.20	.08	2.34	2.81	4.79	1.56
Functional nodal (24)	1.04	−1.50	1.85	2.62	4.12	.44
Government-education (14)	2.50	1.18	3.23	4.37	5.04	1.17
Education-mfg. (5)	1.36	−1.31	1.95	2.34	4.31	1.27
Manufacturing (22)	.63	−1.55	1.64	2.91	3.77	.38
Industrial-mil. (12)	3.14	2.04	3.91	4.12	5.90	1.42
Mining-industrial (12)	1.06	−1.25	2.32	3.50	4.30	1.20
Residential-resort (11)	4.95	4.38	4.90	6.70	6.63	2.42

SOURCE: Compiled from U.S. Department of Labor (1977 and 1983).
NOTE: All 134 SMSAs are among the 140 SMSAs classified by Noyelle and Stanback (1984). Data for 6 SMSAs were not available. Annual rates shown are modified averages (highest and lowest values dropped in each classification, except for national nodal and education-mfg., for which straight averages were computed).
*Numbers in parentheses indicate number of SMSAs.
**The principal components of the "other services" industry group are business services and nonprofit services (largely education and health).

But the aggregate employment data do not tell the whole story. The national nodal, functional nodal, and education-manufacturing have experienced heavy losses in manufacturing employment (see Table 6.2), and have posted aggregate gains only as a result of substantial increases in services, especially FIRE and other services (which include both business and nonprofit services). Moreover, all of the national nodal and many of the functional nodal centers are quite large, so that growth rates fail to reflect the extent of their vitality. Their modest net growth rates represent a considerable measure of success in transforming their economic structures. On the other hand, the manufacturing centers have not only shown little growth but have, for the most part, been unsuccessful in bringing about a transformation toward a service-based economy. Growth rates of their small FIRE and other services sectors are the lowest of all the groups (see Tables 6.2 and 6.3).

Still further insights can be gained by observing the relative shares of employment accounted for by manufacturing and selected service categories and the changes in these shares that have occurred since the mid-1970s (Table 6.3). It is apparent immediately that the nodal centers, residential-resort, education-manufacturing, and government-manufacturing have the most developed service sectors

TABLE 6.3
Average Shares of Employment, Selected Industry Groups, by Type of Metropolitan Economy, 134 Largest SMSAs, 1976 and 1983

Type of SMSA*	Manufacturing		Wholesale-Retail		FIRE		Other Services**		Government	
	1976	1983	1976	1983	1976	1983	1976	1983	1976	1983
National nodal (4)	20.9	17.9	21.0	22.0	8.4	9.6	21.1	25.4	17.0	15.2
Regional nodal (19)	20.7	18.3	24.1	24.6	6.6	7.1	19.6	22.7	16.9	15.3
Subregional nodal (16)	17.7	15.0	24.3	24.6	6.9	7.3	18.1	21.6	19.6	18.4
Functional nodal (24)	32.7	27.5	20.7	21.8	4.5	5.1	17.0	20.9	15.4	14.4
Government-education (14)	16.2	14.5	20.1	21.1	5.1	5.8	17.8	20.9	31.3	28.7
Education-manufacturing (5)	26.7	22.1	20.4	21.1	4.6	5.0	19.2	23.2	21.4	21.3
Manufacturing (22)	37.1	31.9	20.3	21.8	3.8	4.3	15.9	19.5	14.4	14.3
Industrial-military (12)	17.6	16.4	21.4	22.7	4.2	4.5	16.9	20.2	28.9	25.7
Mining-industrial (7)	13.1	12.8	23.4	24.3	4.2	4.6	17.7	21.3	22.7	21.0
Residential-resort (11)	13.3	13.0	25.8	25.6	5.6	6.1	23.2	26.0	20.2	16.8

NOTE: For notes and source, see Table 6.2.

when measured in terms of combined shares of FIRE and other services (which include the financial, business services, and nonprofit services infrastructure) and that the production centers are least well endowed.

In short, the evidence suggests that comparative advantages and disadvantages are deeply rooted and that the weakest places are experiencing little success in effecting the kind of industrial transformation necessary to restore them to economic health.

GROWTH AND SIGNIFICANCE
OF THE NEW TECHNOLOGY

The modern computer traces back to the ENIAC in 1946, but it appears to have first found useful commercial application in the late 1950s. Since its inception there have been a series of breakthroughs in technology that made possible, first, the large high-speed mainframe hardware and, then, successively, the minicomputer and microprocessor, with ever more impressive advancements in computing speed, memory, and cost efficiency in each. There have also been other technologies that have had parallel development and that are increasingly being joined to provide highly versatile and cost efficient new configurations of equipment and applications. These include video, facsimile transmission, television, and, most important, telecommunications.

How are we to begin to understand where these new, powerful technologies are taking us and to assess their likely impact on the American economy, in general, and the American city, in particular? In the view of this nontechnical observer, such understanding requires that we recognize that adoption of computers and related technology has proceeded unevenly, but that virtually every branch of the economy is being impacted by a new era of applications of computer and related technology; that the new technology is bringing about major changes in the way work is done and the way firms are organized; and that these changes are creating new locational options for firms and nonprofit organizations. Finally, it is important to recognize that the special requirements of telecommunications networks in this new era of applications are likely to influence the competitiveness of firms and urban places, creating new opportunities for some and obstacles for others.

UNEVENNESS OF APPLICATION

As the technology has evolved, the extent to which it has been applied has varied widely among users. Such unevenness is due to a number of factors, including the nature of the firm's product (or mission), the size of the user organization, and factors surrounding the investment decisions, such as financial constraints and the progressiveness of management.

In some organizations the new technology is essential because it makes possible certain strategically important service features or because it brings about radical reductions in the cost of production. When this is the case, organizations tend to move quickly to adopt the technology. Thus, when AT&T first offered private-line data communications facilities in 1958 the major airlines promptly installed reservations systems, and stock market quotation facilities began to be put into place only a few years later. Similarly, computer applications have permitted banks to move rapidly toward new forms of competition in which they no longer rely simply on their proximity to retail customers or on long-established ties between their officers and large individual or institutional customers, but increasingly utilize advanced technology to offer variety, flexibility, and speed of transaction in deposit, checking, and portfolio management as well as in borrowing arrangements. Still other examples include insurance firms that seek competitive advantage by using the computer to tailor insurance options and provide more rapid and informed quotation of premium rates, and hotels and rental car agencies that garner customers through automated reservation systems.

Size of user organization also plays a role, of couse, with large organizations tending to be the earliest users in any given industry as well as the most sophisticated. This is true not only because the technology is most effective when applied to the processing of large quantities of data, the handling of heavy telecommunications-oriented traffic, or the need to monitor or coordinate dispersed activities, but also because large organizations are more likely to have the financial means to invest in necessary equipment and software and the staff to dedicate to the planning and supervision involved.

While larger organizations have led the way, smaller organizations have not been barred from making use of the computer. From the outset service bureaus made available the capabilities of large mainframes and the expertise of trained staffs for a variety of fairly routine computational needs such as billing, payroll, and inventory accounting, largely on a batch process basis. However, smaller firms have typically lacked the finances or staff to purchase hardware and software and to create and use more complex data processing systems. In recent years the availability of inexpensive microcomputers along with off-the-shelf software, particularly spread-sheet capability, has opened up opportunities for bringing the computer into the small organization, not only for basic records processing but also for managerial control and planning.

A NEW ERA OF COMPUTERIZATION

What appears to be taking place currently is a rapid transition from an era in which computers were used principally as adjuncts to work done in the various administrative departments to an era in which they are used for a vast array of managerial tasks, including organizational control, planning, and financial and marketing strategies. The new era is one in which the necessary technologies are sufficiently advanced in sophistication and cost effectiveness, and in which software along with professional expertise is sufficiently available, that users are finding themselves under great pressure to adopt the computer or to expand its use. Moreover, it is an era in which the accumulated experience with earlier applications coupled with a new generation of management has ushered in a new environment in which change is likely to be favored rather than resisted. Above all, it is an era in which the several related technologies, especially computers and telecommunications, have been combined to open up entirely new uses, increasingly involving the inauguration of integrated systems and the building of data bases accessible throughout the firm without the earlier restrictions upon space or function.

Richard L. Nolan (1979), a leading computer management consultant, has provided a useful model that sheds light on what is currently taking place by describing how computer processing develops from initial adoption to most advanced application in a prototype large corporation. The 6 stages that he delineates fall into 2 "eras," the "computer era," which includes the stages of (1) initiation, (2) contagion, and (3) control, and the "data resource era," which includes the stages of (4) integration, (5) data administration, and (6) maturity. Progress from stage to stage is driven by the dual effect of accumulating experience within the user organization and rapidly improving technology. During stages 1 and 2 "there is a concentration on labor-intensive automation, scientific support, and clerical replacement," during stages 3 and 4, "applications move out to user locations for data generation and data use," and during the final stages "balance is established between centralized shared data/common systems applications and decentralized user-controlled applications" (Nolan, 1979: 121). The computer era begins with 100% batch processing and moves toward introduction of data base processing, and inquiry and time-sharing processing. The data resource era sees batch and remote job-entry processing progressively reduced and data base and data communications processing becoming the dominant mode (well over 50% of computer usage), and personal computing and minicomputer and microcomputer processing together accounting for over one-fourth.

Nolan's summing up of where the typical large corporation stood at the beginning of the 1980s is as follows:

> [W]e've spent the last 20 years learning how to formally define systems, and install systems analysis programming . . . to provide data processing support to business functions. We've done that with a technology which we generally call the batch technology and, in a real sense, we have faced severe limitations in using that (technology) robustly.
>
> Now what happens in the Stage Three-Four environment is that organizations retrofit those applications . . . to permit them to tap the data resource much more easily. Secondly, organizations bring capability forth to the user (within the organization) through terminal technology, minicomputers and distributed systems. When these two things are fitted together, the results are that 20 years of investment are suddenly unleashed, results that have been pretty well locked up in until-now inflexible systems. These dynamics seem to be what's driving the Stage Four growth that we see. So much latent user demand has been pent up in inflexible systems in earlier years that freeing it causes a near explosion [ICP INTERFACE Data Processing Management, 1980: 19].

There seems to be solid evidence that we are, indeed, entering a new era of computer usage. National Income Division estimates indicate that investment in computer and telecommunications technology—if we define this as office, computer, and accounting machinery, and communications equipment—rose from 18% of total nonresidential durable equipment expenditures in 1976 to 37% in 1982. Still further evidence is found in Nolan's (1979:122) data, which show that data base management software installed on IBM medium- to large-scale computers in the United States rose from about 15% of such equipment in 1975 to roughly 80% in 1980.

In interviews with users, Thierry Noyelle and I have found large banks and insurance companies and several of the large department store chains far advanced in the conversion toward integrated systems, the city of New York moving rapidly toward development of an elaborate telecommunications network and advanced systems of monitoring and administering a variety of the city's services (although by no means all), and relative backwardness in the computerization of administrative procedures of two major universities, although in each plans are underway for major advances in the years just ahead.[3] Even casual reading of either the technical journals or the business press provides additional examples of acceleration and rapidly broadening application of the more sophisticated technology available today.

But the acceleration of application of computers to sophisticated uses is not limited to the installation of distributed processing systems. Among both large users and small the availability of microcomputers at extremely low prices, along with a variety of easily-used software, is setting off a virtual revolution in terms of new applications. In large organizations departments are refusing to wait for the more complex systems and are putting into place subsystems using desk computers to monitor, analyze, and plan their own operations. We have encountered this not only among large firms but also within universities and the various agencies of the city of New York.

In small businesses there has been a rush to apply the new hardware to a variety of uses, such as inventory control, routing and monitoring salesmen, planning and controlling cash flow and investment, filing records, and, of course, word processing.

The applicability of the microprocessor is astonishingly broad and the impact, in terms of efficiency and sophistication of operation, is impressive indeed. This new development raises questions regarding the role of small businesses (both services and manufacturing) and professional organizations. It is entirely possible that large firms, governmental agencies, and nonprofit firms will increasingly find it feasible to contract out additional functions to these smaller organizations and that smaller businesses, given access to telecommunications networks, will be able to expand the geographical scope of their markets.

CHANGING WORK AND ORGANIZATION

The evolution of computerization has involved both a change in the way work is done and a change in functional emphasis within the user firm. On one hand, there has been a change in the work content of occupations and in the relative importance of occupations; on the other, there has been a change in emphasis given to various functions and to the allocation of resources within the user organization.

From the outset computerization made possible the reduction or elimination of some tasks (e.g., filing, laborious computations, and preparation of forms) and created others (e.g., programming, computer operations, and maintenance), while at the same time freeing budgets for a greater emphasis on marketing, product developmnt, and management. As applications have become more sophisticated, and particularly as data base systems have come into use, computers have increasingly been used by management to provide more carefully customized lines of products or services, to control finances, to redesign sales and promotional efforts, and to improve planning and managerial control, all of which are bringing changes in work and in the firms' organization.

Organizational changes have occurred at every level. Distributed processing has permitted the shifting of much of the routine processing out of general computing centers and back to the originating departments. Where high-volume processing such as check clearance, accounts receivable, or credit card processing has continued to require large cadres of clerical personnel, the availability of telecommunication systems has made it possible to relocate away from the headquarters operations to sites where rentals are lower, labor supply is more suitable, and wage rates are more favorable.

In the industries studied there have been marked changes. For the large department store with multibranch operations the installation of point-of-sale terminals has made possible the transmission to higher echelons of the information necessary for the consolidated processing of accounts receivable and billing, and for the closer control of inventory and purchasing. Along with the latter there has come a shift of the buying functions to the central office. With headquarters assuming greater merchandising as well as financial, general policy, and planning functions, the responsibilities of individual store management are increasingly being limited to focusing on selling and general house-keeping.

The large banks and insurance companies have been among the leaders in moving to sophisticated data processing with sharp impacts on work and organization at all levels. Not only did computerized check handling reduce back office staffs at an early stage, but more recently automated teller machines have reduced the use of regular tellers and permitted those who remain to handle a greater variety of transactions. Integrated systems with on-line enquiry have permitted account officers to retrieve information on their customers and to provide a greater range of services. Among insurance companies equally dramatic effects have been observed, with billing and premium accounting becoming automated and centralized, and on-line enquiry into data bases (containing information formerly stored in massive filing systems) making possible more efficient and prompt customer servicing. Problems of rate setting and submission of reports are much more readily solved than formerly, as are problems involving management of investment portfolios, and ready availability of information from central data bases is changing the role of salesmen and the functions of sales offices.

Thus, the general effect has been a redefinition of many jobs, a deletion of some and an addition of others, with some departments taking on new importance and others less. An active controversy is abroad as to whether work is being skilled or deskilled, but the terms of the debate appear to be misstated. While it is clear that computerization has in the past been most successful in eliminating repetitious, low-productivity work, it has also acted to simplify work of all sorts. What seems to be taking place is that, in a rapidly changing environment in which even the lower levels of work often involve the use of expensive equipment and the higher levels involve coping with a much wider range of problems and procedures than formerly (albeit with the powerful assistance of technology), employers are placing a greater emphasis on literacy and ability to learn new ways of work. For higher-level work there is also a new emphasis on credentialing through advanced business or technical training. At the same time the new world of work places greater penalties on those with poor educational backgrounds and levels of competence. Those low-level jobs that are open to them tend to be scheduled on a part-time basis, to pay poorly, and to offer few fringe benefits or opportunities for advancement.

There are also new trends in work arrangements as employers make use of part-time and flex-time workers to accommodate both to the requirements of the new work and the needs of a changing labor force. For the most part, the shift to white-collar work has released employers from the need to follow the old 8-hour staffing patterns of the factory. Through increased use of part-time scheduling to match the flow of work, payroll costs have been reduced in sales and to some extent in clerical occupations. On the labor supply side, there have been major changes in the composition of the labor force. Women have come to play an increasingly important role in the labor force (their share of jobs rising from 29.6% in 1950 to 43% in 1980), and working women are much more likely to be married (59.3% in 1980, compared to 52.1% in 1950) and to have children (62.5% of labor force participants in 1980, up from 28.3% in 1950).[4] This growing importance of women in the workforce has brought new demands for flexible work schedules, more convenient transportation, and better security both on the job and while journeying to work.

TECHNOLOGY AND CHANGING LOCATIONAL OPTIONS

A major implication of these changes in work, organization, and the composition of the labor force is that firms are being faced with new options and new constraints in locating their activities. For the large firm there are opportunities to strengthen some functions at general and divisional headquarters, to thrust some outward to sites where labor is cheaper and of a more suitable quality and where rents and taxes are more favorable, and to eliminate some operations altogether or to consolidate reduced operations in new sites. For smaller organizations, also, there may be new alternatives opening up if data base technology and telecommunications make it possible to service broader markets. On the other hand, cost and supply factors that were acceptable at a given location under previous conditions may erode the firm's competitive position when new procedures and organizational arrangements are put into place.

These new options are not necessarily radical departures from the kind of alternatives that have been available to many users for a number of years, but they are sufficiently different to bring about important changes in the location of service activities over time. Thus, shifts of headquarters and particularly back office activities from city to suburb have been taking place for many years, but the new computer-cum-telecommunications technology makes such relocation feasible for a larger spectrum of activities.

An educated labor supply appears to be emerging as a locational factor of major proportions. The occupational statistics clearly show that there is a rising demand for professional, technical, and managerial staff both in absolute terms and as a relative share of the labor force. Firms and not-for-profit activities must draw upon a pool of such personnel and be able to offer the amenities to hold them. Clerical jobs also continue to increase with the rise of services, but not as a relative share of employment in many industries. The important observation regarding labor requirements for these workers—largely women—is that employers appear to be raising their hiring standards. This may be changing the competitive position of suburbs (where there is typically a pool of educated housewives) relative to the central city.

For many corporate headquarters, R&D, engineering, and related services, proximity to one or more major universities appears to be a decisive factor. This is seen readily in the rapid development of the centers of high tech complexes, where light manufacturing, R&D,

and a variety of engineering and service activities exist cheek by jowl. In the Silicon Valley (near Palo Alto), Route 128 (near Boston), the Research Triangle (North Carolina), and the new high tech complex near Philadelphia, the proximity to major universities is clearly of fundamental locational importance.

But access to a well-developed infrastructure of business and financial and other key supporting services, especially major airport facilities, continues to be a basic requirement for many firms and is the principal source of economic strength of the large nodal center, as the mutual attraction of a variety of headquarters, distribution services, advanced business services, and educational and nonprofit institutions acts to hold existing firms and attract others. The trend toward locating headquarters and related establishments far enough away from the city to take advantage of lower rents, favorable labor supply, and less congested surroundings while remaining close enough to have access to key financial, business, and airline services affects the city's economy, in turn, but just how much and how consistently, from city to city, is a matter about which we as yet know very little.

THE ROLE OF TELECOMMUNICATIONS

Finally, we must recognize the strategic role of telecommunications. Just as the far-flung network of railroads was the infrastructure that permitted the coming of age of industrialization and the relocation of manufacturing and agriculture, so it is that the telecommunications system stands as the basic infrastructure of the new era of computer usage that makes possible the new locational patterns of eonomic activities that are developing.[5]

The most basic telecommunications system for data processing is the existing telephone network, which has been used from the outset. Its principal limitations, however, are that the conducting medium (single copper wire) is limited in capacity and that the American telephone system is only partially equipped with the expensive electronic digital switching equipment necessary for data transmission. Moreover, efficient high-volume data transmission requires not only the availability of such electronic switching, but also the linking of computer hardware and terminals through coaxial cable, microwave radio (either satellite or terrestrial), or fiber optic cable.

The requirement of a highly capital-intensive infrastructure creates significant inequalities of opportunity among users and among places in terms of their ability to make use of integrated systems over broad geographical areas. A number of very large corporations (e.g., Citibank, Sears, J.C. Penny) have invested enormous sums to put into place nationwide and even international systems linking their various establishments, but smaller users or users with less extensive requirements must be able to tie into established networks in order to make use of distributed data processing. For them, the facilities of the local telephone system may be the limiting factor, as lack of electronic switching, which is available for well under 50% of the national network, may foreclose extended use of sophisticated data processing systems in many of the smaller places for at least some years to come.[6]

Sophisticated telecommunications systems, like modern jet ports, may be expected to develop most rapidly where traffic is heaviest, thereby reinforcing initial comparative advantages of those places that are already the most advanced as service centers or that are developing most rapidly and are likely to attract the earliest capital outlays. Thus, New York City, which is the largest and most important service center of the nation, is equipped with the most sophisticated telecommunications system. Its telephone system has been substantially modernized and, for the entire business district, at least, converted to electronic switching. Manhattan is ringed by an optical fiber system and fairly crisscrossed with coaxial systems. Several satellite stations either are located within the city or are available to users through coaxial or fiber optic cable.[7] It is difficult to imagine that such facilities do not provide a major source of strength to the city's economy in enabling it to hold onto its position of preeminence in finance and commerce.

But telecommunications infrastructure requirements do not appear to offer the same impediments to dispersal of activities to suburban locations of large cities as to distant smaller cities or towns. Telephone networks are likely to be upgraded earlier in large metropolitan areas than in smaller places at more distant points, and alternate systems using more advanced technology and linking the suburbs with the central business district are becoming increasingly available throughout the larger metropolitan areas.

TECHNOLOGY AND COMPETITION WITHIN THE AMERICAN SYSTEM OF METROPOLITAN PLACES

A principal conclusion of the preceding analysis is that the accelerating application of computer and related technologies along with the rise of services is bringing about new possibilities for location of firms and leading to a greater competition among places for growth and development at two levels: among metropolitan areas and between cities and suburbs within given metropolitan areas. In each arena of competition the outcome will be influenced by locational attractions of labor force quality, availability of telecommunications networks, the infrastructure of business services, educational, government, and nonprofit institutions, and a number of quality-of-life factors.

The argument sometimes put forward that the slow-growth places, and particularly old industrial centers under stress, can be rescued by bringing in new service activities by means of the computer and telecommunications technology must be carefully reevaluated in light of this discussion. For these places the prerequisites of an appropriate labor supply, a minimum complex of supportive services, and adequate telecommunications are likely to be lacking.

It may well be possible to revitalize these places, but radical therapy probably will be called for. Their comparative advantages lie, for the most part, in the area of production. Large infusions of capital to bring in new manufacturing technology could, at least in some places, reestablish a viable export base around which supporting service activities could flourish.

But there is a more general lesson to be learned regarding the implications of the new technology for urban growth and development. This lesson, to be drawn from all that has been said above, is that the new applications of the computer and its increasing adoption in all types of economic activity, private and not-for-profit sectors alike, place new demands on the labor force and the urban infrastructure. Regardless of what the data suggest about the comparative advantages of certain types of places, these new demands call for new initiatives in all places that seek to maximize economic well-being. What is called for universally is an upgrading of educational and other training institutions (including new arrangements for retraining workers at all stages of their careers), improved counseling and employment services, improvement in transportation services and in public sector services relating to safety and health and to the care of the children of working mothers, and a concerted effort to affect the quality of life in such a way as to hold onto and attract establishments that are, in many instances, being faced with a much broader set of locational options than heretofore.

This general conclusion would appear to be particularly relevant for cities and suburban communities that must face the issue of developmental strategies under conditions in which the new technology is raising new choices of location for firms within the metropolitan area. It is not yet clear whether the new locational trends will, on balance, favor city or suburb. What does seem clear is that the matter is not altogether in the hands of the gods. Appropriate and aggressive public policy can do much to strengthen the local economy.

NOTES

1. For a more extended analysis of the rise of services, see Stanback et al. (1981).

2. The rising importance of services has by no means been restricted to the postwar years. During the period dating roughly from the last quarter of the nineteenth century until the Great Depression, the rise of the industrial economy was accompanied, and in many ways facilitated, by a rapid rise in distributive and in certain producers' services, such as banking and insurance. The postwar period has brought a different pattern of changes, however, and is the only concern here.

3. For an enlightening and entertaining summary of the New York City experience, see Smith (1984).

4. The percentages of women with children are not strictly comparable because of data reporting changes, but they do provide a rough approximation of the changes involved (see Dutka, 1984).

5. Gershung and Miles (1983) have called attention to the importance of the telecommunications infrastructure in their recent analysis of the new services economy in the European economy.

6. Distributed data processing systems can, of course, be established by large organizations for local areas using telephone networks and privately owned switching equipment.

7. For a more detailed description of New York's telecommunication system, see Moss (1984).

REFERENCES

DUTKA, A. B. (1984) New Types of Work Scheduling: The United States' Experience. New York: Columbia University, Conservation of Human Resources Project.

GERSHUNG, J. I. and I. D. MILES (1983) The New Service Economy: The Transformation of Employment in Industrial Societies. New York: Praeger.

HARRIS, C. S. (1983) Small Business and Job Generation: A Changing Economy or Differing Methodologies? Washington, DC: U.S. Small Business Administration.

ICP INTERFACE Data Processing Management (1980) "ICP interview: Richard L. Nolan." (Summer): 19.

MOSS, M. L. (1984) "New telecommunication technologies and regional development." Report to the Port Authority of New York and New Jersey. (unpublished)

NOLAN, R. L. (1979) "Managing the crisis in data processing." Harvard Business Review 58 (March-April): 115-126.

NOYELLE, T. J. and T. M. STANBACK, Jr. (1984) The Economic Transformation of American Cities. Totowa, NJ: Rowman and Allanheld.

SMITH, D. (1984) "Brave new city government." New York Magazine 17 (May 14): 56-64.

STANBACK, T. M., Jr., P. J. BEARSE, T. J. NOYELLE, and R. A. KARASEK (1981) Services/The New Economy. Totowa, NJ: Allanheld, Osmun.

U.S. Department of Labor (June, 1983) Employment and Earnings. Washington, DC: author.

—(June, 1977) Employment and Earnings. Washington, DC: author.

7

Office Automation and Women's Work: The Technological Transformation of the Insurance Industry

BARBARA BARAN

□ NOT ONLY is the majority of our work force now engaged in various forms of white-collar activity, most of it in offices, office work is also largely women's work. Of the total female work force, 67% is employed in white-collar occupations, compared to only 40% of the male work force. More than one of every three working women is a clerical (U.S. Department of Labor, 1981). Today, the introduction of computer-based technologies into offices promises to affect dramatically the working lives of these women. Although the "office revolution" has been much heralded but long awaited, this chapter will contend that in the last few years the combined forces of heightened competition and technological innovation have sped the process of diffusion. Automation of white-collar activities is proceeding much more slowly than its enthusiasts had hoped; nevertheless, in a number of industries information technologies are beginning to have a major impact on the size of the work force and the nature of work.

AUTHOR'S NOTE: *Research for this chapter was done in conjunction with Suzanne Teegarden, Barbara Facher, and Jean Ross. In addition, I would like to thank Stephen Cohen, Ben Harrison, Ann Markusen, Michael Teitz, and, of course, Manuel Castells for their help along the way, and Eileen Appelbaum for sharing her own work on the insurance industry. Finally, I am especially indebted to three loyal critics and supporters: Amy Glasmeier, Margaret Baran, and, as always, Jim Shoch. A portion of this research was performed under contract with the United States Congress Office of Technology Assessment through the Berkeley Roundtable on the International Economy; however, the opinions expressed here are mine.*

To date, however, there has been little empirical analysis of these changes. Most of the literature on the effects of technology on work design and manpower requirements has been, and remains, focused on manufacturing. In addition, insofar as a literature on office automation does exist, it is remarkable for the disagreement among researchers.

The intent of my research on the insurance industry, then, was to contribute to this nascent literature—with a particular emphasis on the probable impacts of office automation on the female work force. I selected the insurance industry for observation for several reasons. First, it is a prototypical white-collar industry in the sense that the production of its product, as well as the supporting administrative activity, involves the processing and generation of paper. Second, insurance is a major employer of office workers, especially clericals; of the more than 14 million "administrative support personnel" employed in 1980, close to 1 million were working in this one industry alone (U.S. Census of the Population, 1980). As a result, of course, approximately 60% of the work force is female (U.S. Department of Labor, various years: percentages computed by comparing Tables B-3 and B-2). Finally, insurance is also one of the most highly automated of the white-collar industries; for example, it has been estimated that insurance companies account for a remarkable 16% of IBM's entire installed base (interview with executive in the Life Office Management Association).

Because aggregate data are both woefully incomplete and, at best, fail to capture with any richness changes occurring in a labor process, I chose a methodology that combined case work and elite interviewing. Initially, a case study was conducted of a major national property/casualty insurer that included 26 interviews with employees in various parts of the company's operations (home office, branch office, data processing center, commercial group, and personal lines center) and analysis of extensive personnel data that the company made available. The second stage of the study involved interviews with executives, personnel managers, and systems analysts in 25 other companies, loosely stratified by size, product type, growth rate, distribution system, and so on. Finally, members of the industry's trade associations, agents' associations, and vendor companies were also interviewed. This fieldwork has been supplemented with secondary source material from government agencies, trade publications, and documents and surveys kindly provided by the trade associations and consultants to the industry.

This chapter begins with a brief review of the existing literature as the context for my own observations. In the following two sections I then argue that, in the insurance industry at least, current and future applications of computer technologies are likely to differ significantly from earlier implementations of office automation equipment (with which most of the literature is concerned); this is because of important changes that are occurring in the competitive environment, in markets, products, and in the technology itself. Finally, I will explore the likely effects of the new implementations on manpower requirements, the occupational structure, and women's labor.

THE DESKILLING DEBATE

Although the deskilling debate certainly preceded Braverman, his contribution was particularly provocative and therefore formed the starting point of this investigation. Braverman (1974) argues that despite the differences between the old and the new technologies, the dynamic of the transformation is a familiar one. Just as the Industrial Revolution robbed skilled artisans of their trade, automation of the office will deskill and degrade white-collar labor.

As in the factory, office work is being divided into hundreds of minute fragments parcelled out among thousands of "detail" workers who thereby lose control over the product of their labor and any variety in their daily activity. The greatest portion of every day, or perhaps the entire day, is spent operating a machine. As middle-level jobs are deskilled, the work force is polarizing into a mass of unskilled workers on one hand and a tiny coterie of managers and professionals on the other.

In recent years the most important challenge to this perspective has come from a disparate group of writers whom I will label the "postindustrialists." I include within this category analysts of diametrically opposed ideological persuasion, such as Daniel Bell (1976) on one hand, and Fred Block and Larry Hirschhorn (1979) on the other, and futurists such as Alvin Toffler (1980).

The common core of the postindustrialists' argument is that the inherent logic of an information-based economy differs in fundamental respects from the first wave of mechanization. First, whereas the assembly line socialized and centralized production, the new technologies offer the possibility of radical decentralization. Second, whereas the Industrial Revolution intensified the division of labor, computer technologies promise less functional specialization. And, third, whereas partial mechanization produced degraded job configurations, fully automated labor processes expel unskilled labor;

within the capital-intensive sectors, then, average skill levels will rise.

In the most optimistic and popular of these accounts, the end product of this transformation is a society of knowledge-workers. Hirschhorn and others, however, warn that so long as outmoded social relations confine the direction of development, we are instead likely to produce a society characterized by serious underutilization of workers' skills and widespread permanent structural unemployment.

Finally, the numerous feminist writers who have intervened in this debate (see, for example, Barker and Downing, 1980; Davies, 1975; Glenn and Feldberg, 1977, 1983; Hacker, 1979; Nussbaum and Gregory, 1980; Werneke, 1983; West, 1982) argue that, because of the pervasiveness of occupational sex segregation within the office work force, women will bear the brunt of the restructuring process. Whereas women's jobs are disproportionately disappearing and their opportunities for upward mobility declining, men may actually benefit both because they will dominate the more highly skilled technical and professional jobs being created and because automation may centralize control in the hands of (male) senior managers and systems analysts.

The limited number of case studies that have been published report conflicting findings. In regard to changes in the occupational structure, some researchers have found that job loss is concentrated among low-skilled clericals (Faunce et al., 1962; Shepherd, 1971), implying a general upgrading of labor, other studies indicate, on the contrary, the elimination of skilled clerical activities, resulting in a polarization of the occupational structure (Glenn and Feldberg, 1977; Hoos, 1961; BLS Bulletin #1468). Similarly, whereas some analysts have reported less task fragmentation as the technology becomes more sophisticated (Shepherd, 1971; Matteis, 1979; Sirbu, 1982; Adler, 1983; Appelbaum, 1984), others suggest that job content is narrowed and worker autonomy reduced (Murphree, 1982; Greenbaum, 1979; Cummings, 1977; Glenn and Feldberg, 1983, 1977). Finally, although in all cases women experienced the greatest job loss, in some reports it appeared that after automation women were relegated to lower-skilled activities (Glenn and Feldberg, 1977; Murphree, 1982), whereas in other accounts female clericals seemed to benefit from the new labor process (Matteis, 1979; Cummings, 1977).

This chapter will address a number of these questions based on my empirical observations in the insurance industry; briefly, however, my critiques of the dominant perspectives can be summarized as follows.

First, although the strength of the Bravermanist account lies in its insistence on an enduring capitalist logic that constrains the implementation of new technologies, insofar as this analysis elevates the factory—with its assembly line and detail workers—to an essential aspect of the capitalist organization of production, the Bravermanists miss much of what is radically new about the new technologies. The leading postindustrialists and futurists, on the other hand, make the reverse error. Bell (1976), for example, seems to have implicitly assumed that the U.S. economy would escape the unpleasant consequences of serious competition, permitting a friendly accommodation between capital and labor. Toffler (1980) similarly ignores the implications of capital's current drive to reduce costs, particularly labor costs, and the extent to which control over the labor force looms large in managerial choice of appropriate technologies, job design, and so on.

In terms of specific analyses of the effects of office automation, the feminist work has, from my perspective, been the most interesting, although, unfortunately, these investigations have relied almost exclusively on the Bravermanist framework. Nonetheless, the feminists have correctly insisted that in the face of the historic and extensive sex segregation of the office, analyses of the likely effects of automation on this work force can ill afford to be gender blind; indeed, these analyses must address the broader questions of changes that are occurring in gender relations and in the role of female labor in the economy.

Finally, however, I will argue that office automation cannot be analyzed as a unitary phenomenon. Impacts vary on the basis of the specific kind of technology introduced, the nature of the work being automated (originating in the unique characteristics of the product, markets, organizational structure, and competitive dynamics of the industry), and perhaps also the characteristics of the available work force. Similarly, it is necessary to periodize the process of office automation. The first wave did, indeed, correspond to the expectations of the Bravermanists. Nevertheless, in the last few years it has been possible to discern the outlines of a fundamentally new dynamic that I will attempt to describe here.

THE FIRST WAVE

The first applications of computer-based technology in the insurance industry involved simple mechanization of extremely structured, high-volume operations. The early mainframes were used primarily as numbers-crunching machines. The automated tasks were fairly discrete so that the shift from manual to computerized performance had little effect on the organization as a whole; task fragmentation, or Taylorization, had already isolated these routinized functions. Word processors, duplicating machines, private branch exchanges, and so on similarly were introduced to mechanize a particular task without regard for the nature of the overall office procedure.

Both the cost and technical requirements of the early machines and the tendency to automate in conformity with the rationalized structure of traditional administrative bureaucracies resulted in intensification of the long-term trend toward task fragmentation and functional centralization. Conceptually and in practice, at this stage of development, Taylorization and mechanization were integrated processes.

Routine keyboarding was separated more sharply from other clerical functions and was often spatially isolated from the rest of the firm. Work in these processing centers (both data processing and word processing) was machine-linked, machine-paced, and often machine-supervised.

Task fragmentation was not limited to processing functions. Overall, in the last two decades the labor process in insurance companies was centralized by narrow function. For example, in a number of companies even the main professional occupation, underwriting, was rationalized: At the bottom end, low-level underwriting functions were transferred to a newly created clerical position; at the top end, more specialty underwriting categories were created. Workers were grouped into sections of underwriters, raters, typists; paper flowed from one section to another mimicking the assembly line. This spatial segregation by occupation was tantamount to spatial segregation by gender. The professional and managerial categories were overwhelmingly male; the growing clerical work force solidly and increasingly female.

In line with the Bravermanist model, then, the first stage of automation in the insurance industry tended to increase job fragmentation, centralize production by narrow function, heighten occupational sex segregation, and make many routine keyboarding functions spatially "footloose." More recently, however, the greater sophistication of the technologies and transformed market conditions are dictating a new organizational logic that promises to reverse many of these early trends.

THE OFFICE OF THE FUTURE

COMPETITION AND ORGANIZATIONAL CHANGE IN THE INSURANCE INDUSTRY

During the 1970s the implementation of advanced office systems proceeded much more slowly than observers and especially the vendors had expected. The barriers to rapid diffusion were both technological and organizational. First, at the level of the technology itself, the task of representing more complex white-collar activities in computer algorithms proved to be exceedingly difficult. To paraphrase systems analyst West Churchman, it was not so easy to get a machine to behave as competently as a clerk (Mintzberg, 1972). The attempt to impose a standardized logic on unstructured procedures generated too many exceptions to be cost-effective and in some cases even diminished the efficiency of the organization (Strassman, 1980). In addition, competition within the vendor community slowed the development of the effective communications networks that are critical to the implementation of sophisticated, integrated office systems.

Second, however, resistance to automation within the user community proved to be equally serious barrier to diffusion of the new technologies. Professionals, managers, and even executive secretaries balked at the introduction of machinery that threatened to transform the character of their work and the relations of power in their workplaces. In the past few years it has taken a dramatic change in the competitive dynamics of the industry to begin to erode this resistance.

Falling interest rates and the slow but steady deregulation of the financial services sector has brought new competitors into the game and cut sharply into insurers' profit margins. Whereas formerly, as a regulated industry, insurance carriers were virtually assured a reasonable rate of return, today even some giants in the field are struggling for survival. Companies in which paternalism protected job security even through the Great Depression have recently been rocked to their foundation by layoffs of up to 1500 employees virtually overnight.

TECHNOLOGICAL CHANGE IN THE INSURANCE INDUSTRY

At the same time, three technical developments in particular have improved the cost-effectiveness of systems implementation and widened the possible range of applications. First, the increasing miniaturization of electronic circuits has produced quantum leaps in the computing power of ever smaller and cheaper machines. Second, significant improvements in telecommunications technology and the merging of communications and computer technologies have allowed insurers to link their data processing equipment to numerous other office machines (such as word processers, fascimile transmitters, microfilmers, and optical character recognition devices) both within buildings and across continents. Third, the simplification of computer language and greater sophistication of computer software has made possible a whole new range of applications.

In contrast to many office environments, systems development in the insurance industry has been propelled by data processing needs. In the last decade mainframes have become increasingly fast and storage capacity has grown geometrically; nevertheless, the most important development in data processing hardware has probably been the dramatic increase in the capabilities of minicomputers and now microcomputers.

Together with the improvements in communications technologies, these sophisticated small computers have allowed insurers to develop integrated, decentralized data processing systems. Such systems typically link a single host computer to a node, which might be a cathrode ray terminal (CRT), intelligent terminal, mini or microcomputer, printer, teleprinting terminal, and so on; the network may be either local or long distance. Local area networks connect a number of components at the same site, often by means of a coaxial cable. Long distance networks perform the same function for remote sites; data transmission is usually through ordinary telephone circuits or leased lines, although some companies have begun to experiment with satellite transmission.

On the basis of these integrated systems insurers have been able to consolidate all policyholder data into central master records stored in the company's main computer installation; in the past, these records were duplicated in up to a dozen functional units. Now users in remote sites can access and manipulate the same data base. As a result, both data entry and data processing functions are often distributed and decentralized; and given the improvements in both mainframe mass storage capacity and communications networks, these decentralized systems are moving rapidly to increase the numbers of on-line, real-time applications.

The development of more sophisticated software is also allowing companies to automate complex, less structured functions; and the new "user friendly" languages are encouraging professionals and managers to directly access and make use of the wealth of information that the integrated data bases make available for strategic planning, marketing, and so on. As a result, true management information or decision support systems are beginning to be possible and professional functions are being automated for the first time.

A NEW APPROACH TO AUTOMATION

Beginning in the late 1970s, then, as the range of possible computer implementations widened and systems analysts began winning their battle against middle-level knowledge workers, approaches to the problem of automation changed as dramatically as the machines. This evolution can be characterized as the movement from functional approaches to systems approaches; that is, from automating discrete tasks (e.g., typing, calculating) to rationalizing an entire procedure (e.g., new business issuance, claims processing) to restructuring and integrating all the procedures involved in a particular division, product line, or group of product lines.

The transition to a systems approach to automation began, as we might expect, with the simplest procedures—procedures that involved a very small number of discrete tasks, such as payroll, accounting, billing, and so on. Over the last decade more and more complex procedures have been automated and the attention of managers and systems analysts in the industry has turned to the problem of a rational reorganization of the entire firm, including its distribution system, on the basis of fully integrated computer systems. All the work involved in a particular product line is analyzed; the labor process is restructured and, where possible, automated. But automation follows the new logic of the organization; older functions may simply be eliminated as artifacts of an outmoded production process. New business issuance is a good example.

The trend in new business processing has been toward on-line, single-source entry. In a great many of these applications, rating, billing, and printing functions are performed entirely by the computer. The underwriting function is computer-assisted; that is, the computer provides actuarial guidelines, performs calculations, facilitates rapid access to policyholder files, and so on. Increasingly the policy moves electronically through the various steps in this process, producing daily reports on policy status at each work station. In the case of highly standarized policies, the computer may even perform the underwriting function.

Most of these systems are based on distributed processing networks that give remote site users access to the company's master records. Often they also reflect the trend toward consolidation, integrating functions such as rate quoting, billing, and calculation of commissions with the new business production.

On the basis of these integrated, on-line systems, insurance companies have recently begun a serious redesign of the labor process. Although Taylorist logic continues to inform their efforts, the emerging organization of work bears little resemblance to the assembly line.

THE REDESIGN OF WORK

The ideal operative on Ford's assembly line turned one bolt. The ideal production worker in the "insurance company of the future," with the aid of a sophisticated computerized work station, rates, underwrites, and issues all new policies for some subset of the company's customers, handles the updates and renewals on those policies, and, as a by-product, enters the information necessary for the automatic generation of management reports, actuarial decisions, and so on.

In many companies studied two closely related processes are occuring. The first is the emergence of highly computer-linked, multiactivity jobs that combine tasks formerly performed by data entry clerks, other clerical workers, and professionals. Typically, data entry, rating, routine underwriting, and sometimes policy preparation have been transferred to a skilled clerical worker; a similar configuration has developed in the claims handling process.

In these cases, unskilled clerical work has been largely eliminated; workers in the new clerical jobs, although closely circumscribed in their decision making, are a long way from the typing pool. Judgment calls do have to be made by these clerks. Often they are required to interact directly with the agents, a level of responsibility formerly reserved for professional underwriters. But, perhaps most important, because these workers are accepting risk and directly issuing policies, they are almost solely responsible for "quality control." Some kinds of quality checks are, of course, built into the computer systems, but at this stage at least such monitoring only functions to catch gross errors and inconsistencies. Basically, in increasing numbers of insurance lines, clerks are reponsible for the soundness and accuracy of the millions of routine risks their companies write and claims they settle.

In these automated processes professionals are also having their work redefined as clerks take over their lower-level functions. First, they have become "exceptions" handlers; that is, they are responsible only for the policies or claims that fail to fit into the "pigeon-holes" of the computerized system (which, of course, includes large risks); therefore, their work has become more interesting and complex. Second, in the case of underwriters at least, there has been a reorientation of the job function away from churning out policies and toward planning and marketing. Increasingly, underwriters are being required to go into the field and work directly with agents to develop, evaluate, and encourage the sale of new products. In this way their job has been enlarged or reintegrated as it combines tasks formerly divided among managerial, professional, and sales personnel; "mental" and "manual" labor are also to some extent being recombined as higher-level employees are required to enter the data directly into the machines, eliminating numerous clerical intermediaries.

The second major reintegrative process that is occurring is the elimination of single-activity units in favor of multiactivity teams. Whereas formerly typists, raters, and underwriters were divided into separate units, each with its own supervisor, now a small team consisting of one or two of each kind of worker will service some subset (often geographical) of the company's customers. In some cases, this form of organization probably simply presages and prepares the organization for the more complete electronic reintegration of work just described. In other cases, however, where the products are more specialized, complex, low volume, or changing rapidly, the team structure is likely to be a long-term arrangement.

Significantly, even the physical environment reflects this new approach. The walls of private offices have been torn down and managers, professionals, and clericals work side by side in shoulder-high cubicles. Open office plans not only reduce overhead costs and permit flexible response to changing technologies, they work to erode outmoded social relations.

Paul Strassman (1980) of the Xerox corporation, has argued that one of the most serious barriers to the improvement of white-collar productivity is "the view [of managers and professionals] that information is a private good, a source of [their] power." In order for information flows to be managed in a conscious and explicit way as the new technologies require, the old power bases have to be destroyed. The new office designs have the dual effect of permitting visual scrutiny of everyone's activities and of removing an important symbol of the old managerial role in which power was based on secret information and private loyalties. The pooling of secretaries has a similar impact.

Finally, however, as Sabel (1982) has suggested in the case of manufacturing, the trends just outlined are translating into different actual job configurations, depending on the nature of both the product and its market. The insurance industry, like many other U.S. industries, has been moving in the direction of more sophisticated market segmentation. As one result, a bifurcation seems to be occurring in their product offerings between highly standardized products aimed at mass markets on one hand, and a proliferation of specialized products, many directed at the upscale market of households with yearly incomes of $50,000 or more.

Both developments are being spurred by the new technologies, but the production process varies significantly in these two situations. High-volume insurance products—such as homeowners packages, health, and auto—are being increasingly standardized so that their production and/or servicing can be almost fully automated. In numbers of these lines, low-level and high-level labor is being expelled, leaving primarily skilled clerks and a few managers. In others, the occupational structure is bifurcating; the computerized system is able to handle all but the most complex decisions so that only data entry clerks and exceptions underwriters are needed to process the policies. Again, the nature of the product seems to be critical in determining which of these two configurations emerges, although factors such as the importance of quality control and the prejudices of top management also play an important role.

At the other end of the spectrum, many of the new specialty and financial services products are too complex, unique, or flexible to lend themselves to this kind of automation. Much of the labor force is professional and is more highly skilled than in the past, as evidenced in some cases by new training and licensing requirements. Tailored products could not be sold profitably without the use of the computer, but here the machine augments rather than assumes the functions of professional labor. Routine clerical functions are also less automated, although they are decentralized in the manner described earlier to reduce redundancy, and the trend is toward elimination of as many manual tasks as possible.

SPATIAL REORGANIZATION OF WORK

Predictably, a transformation of the industry's spatial organization is accompanying this redesign of the labor process. The direction of change is not a simple one of more or less centralization; several driving principles are operative producing a variety of configurations. Perhaps the two most important motivations are the desire to eliminate expensive redundancies and, more generally, a push to reduce land, labor, and overhead costs (although companies are also shifting work to improve customer service, data quality, and the sales effort). Small powerful computers and expanding telecommunications capabilities have opened up numbers of new ways to meet these goals. Companies no longer have to locate whole procedures in the same site; the various component functions can be performed in spatially disparate locations and then reintegrated electronically.

For example, to eliminate redundancy and take full advantage of scale economies, high-volume, standardized processing is being centralized for an entire multilocational corporation. In the case of insurance fleets (that is, combinations of numbers of separate companies), one data center sometimes services the entire fleet. Because the occupational composition of these centers is largely technical and clerical, they are usually located away from the expensive land and labor of the central cities, but in areas where telecommunications hookups are adequate.

At the other extreme, the drive toward greater efficiency has led insurers to decentralize numbers of functions: Data entry and often certain kinds of printing and processing functions are being pushed down to the point of data generation, in this case often the agency or field office. Single-source entry and decentralized production eliminate innumerable redundant steps, improve the accuracy of the data, and dramatically speed turn-around time.

In some cases even more radical decentralization is occurring. Although still rare, insurance companies are beginning to experiment with telecommuting, or homework. The motivation is both cost reduction and the desire to attract or retain the preferred work force—educated, preferably married, and usually white women.

One company studied now has about one-sixth of their claims adjustors working at home. These women (and they are all women) operate as independent contractors and are paid on a piece-rate basis. Using interactive terminals linked to the companys mainframe, they can log onto the system at their convenience; the computer stacks the claims in a que, automatically updates the master file, creates the checks and explanations of benefits. Productivity is extremely high; overhead and labor costs, low. Although both the information and physical production remain centralized, the labor process itself has been decentralized, decollectivized, and rendered asynchronous.

Finally, however, whole procedures are also being relocated. For example, in order to automate an entire product line, many companies have consolidated work formerly scattered in numerous branch offices into two or three regional centers. Often there is no longer any reason that production has to occur in proximity to the customer base.

SUMMARY

In all of these ways the emerging labor process in the insurance industry is the antithesis of the traditional factory. Electronically reintegrated job categories are combining numerous narrow occupations; in a very real sense, the functioning of the assembly line is now internal to the machine. Similarly, whereas the factory socialized production, the new technologies are paving the way for an organization of work that is extraordinarily isolating.

In the new offices more and more categories of employees now sit alone in small cubicles riveted to their terminals. In many companies the ratio of terminals to employees is already 1:2. With the move to interactive, on-line systems, these workers are increasingly engaged in "conversations" with the machine rather than with coworkers.

In the most advanced systems there is also little immediate, physical interdependence among workers—either because the various steps in the production process can be separated in both time and space or because one clerk assumes all the functions of the production team. The current experiments with home work are the radical extreme—and perhaps the logical extension—of this process.

REDUNDANCY, SKILL, AND THE NEW EXPERIENCE OF WORK

On the basis of these observations, then, information technologies are likely to have two major impacts on the manpower requirements and occupational structure of the insurance industry over the next decade. First, employment should fall, at least in relation to output and probably in absolute terms. Second, skill levels should rise as routine clerical work is eliminated and professional work becomes more complex. At the same time, however, opportunities for occupational mobility will probably decline, the quality of work life may deteriorate, and workers may have less discretion and independence in the conduct of their jobs.

THE ELIMINATION OF WORK

Although true productivity figures are almost impossible to calculate for the insurance industry,[1] there is considerable evidence that automation is improving efficiency and reducing labor inputs. In the life segment, where automation of policy production has proceeded most rapidly in the last decade, purchases of life policies

TABLE 7.1
Percentage Distribution of Insurance
Employment by Occupation: 1960, 1970, 1978

Occupation	Percentage Employment			Percentage Change
	1960	1970	1978	1970-1978
Managers/officers	13.3	11.8	12.1	21.4
Prof/Tech workers	3.2	5.8	6.0	24.0
Clerical workers	47.4	50.0	45.4	8.0
computer operator		0.7	1.3	119.0
keypunch operator		1.8	1.2	−21.9
statistical clerk		2.8	2.5	6.4
bookkeeper		3.6	2.8	−7.4
adjustor/examiner		7.3	9.8	58.8
file clerk		2.5	1.7	−19.7
mail handler		0.7	0.5	−10.7
secretary		13.2	12.0	8.2
typist		6.9	5.1	−11.7
Sales	33.6	30.0	34.5	37.2

SOURCE: U.S. Department of Labor (1969 and 1981).

increased by 49.2% between 1970 and 1980 while the labor force expanded by only 9.8%. Between 1980 and 1982 sales grew by 20.8% and employment remained essentially stable.[2]

The situation in the property-casualty segment is more complicated for several reasons: fewer lines of business have been automated, greater reliance on an independent agency force makes it more difficult for insurers to develop fully integrated systems, and for the last decade companies have been scrambling to expand sales to reap the investment benefits of high interest rates.[3] As a result, between 1970 and 1980 employment kept pace with gains in output.

Recently, however, reckless expansion caught up with these companies; losses and layoffs have been widespread. Between 1980 and 1982 sales plunged by 6.5% and employment by 3.8%.[4] Since then labor-shedding has probably increased. With one exception, every property-casualty company visited had experienced serious layoffs in the last 4 years, ranging between 5% and 15% of their work force. None of these companies foresee new hiring. Heightened competition has brought renewed concern with the bottom line; property-casualty insurers are moving rapidly to automate new product lines and develop an industrywide agency value-added network (with IBM) to improve the efficiency of their distribution system.

Although the layoffs in response to overcapacity were felt throughout the occupational hierarchy, the impacts of technological

TABLE 7.2
Percentage Change in Employment in Insurance Carriers
by Selected Occupation: 1978-1981

Occupation	Percentage Change in Employment
Total employment	5.6
Managers and officers	10.9
Professional and technical	9.5
actuary	21.7
systems analyst	42.5
accountant/auditor	−51.1
claims examiner, P/C	10.8
underwriter	3.0
computer programmer	36.2
Clerical	3.5
computer operator	0.4
bookkeeper, hand	−15.2
claims adjuster	−7.7
correspondence clerk	−1.5
file clerk	−15.9
general clerk	−5.9
rater	−1.6
secretary	6.6
stenographer	−31.4
typist	−2.9
Sales	6.6

SOURCE: U.S. Department of Labor (1981/1978).

change have been more discriminate.[5] Routine clerical categories were the first to be eliminated and, for the next few years at least, will continue to be the hardest hit. Between 1970 and 1978 clerical employment as a percentage of total employment in the industry fell from 50% to 45%; in effect, 73,000 clerical jobs were eliminated by the shift in the occupational structure. Keyboarding and filing occupations were particularly seriously affected; the number of keypunch operators dropped by 22%, the number of file clerks by 20%, and the number of typists by 12% (see Table 7.1).

Other categories of clerks also experienced absolute declines in employment. For example, between 1978 and 1981 the number of stenographers fell by 31% and the number of manual bookkeepers by 15%. Because of the newly automated claims and underwriting support systems, in the space of these 3 years the industry also employed fewer claims adjustors and raters (see Table 7.2).

Since 1978 also, the impacts of automation on professional work are beginning to be visible in the aggregate data. Between 1978 and

1981 underwriters lost employment share and the number of accountants and auditors plunged dramatically. Although the professional and technical category continued to grow faster than total employment, most of this expansion was directly attributable to the rapid addition of computer professionals (systems analysts and programmers); excluding these two categories, professional occupations grew at about half the rate of total employment (see Table 7.2). Because computerization of professional underwriting and claims functions did not begin in earnest until the end of the last decade, we should expect these trends to accelerate.

CHANGES IN SKILL REQUIREMENTS
AND THE OCCUPATIONAL STRUCTURE

In contrast to analysts, then, who argue that automation tends to eliminate higher- and middle-level clerical occupations, my own work suggests that most traditional clerical categories are declining—routine, semiskilled, and skilled—including fairly new, but now obsolete, computer-linked jobs such as keypunch. At the same time, however, because computers are best able to perform highly structured functions, narrow routine clerical jobs are disappearing the most rapidly. Whereas the first round of automation did, indeed, deskill and "proletarianize" much clerical work, the second round is totally eliminating many of those degraded functions.

What is being deskilled, on the other hand, is professional work. Today both users and vendors of office automation equipment have largely shifted their attention away from the task of automating routine activities and are struggling with the problems of how to translate the functions performed by knowledge-workers into computer software. As they are successful these functions are turned over to cheaper labor.

Ironically, in net, the rolling process of deskilling and redundancy is probably raising average skill levels within the industry. Large numbers of low-skilled clerks who formerly make up the bulk of the work force are being expelled; skilled clericals are taking over low-level professional functions; and (a smaller) professional staff is having its work upgraded. In all categories, then, there are relatively fewer jobs, but the remaining work is, overall, more skilled. In this case, therefore, it is critical conceptually to separate the individual workers from their functions. Elements of the work process are being degraded, but the specific agents now performing those functions are either being upgraded or rendered redundant.

OCCUPATIONAL MOBILITY

Despite rising skill levels in the industry, career paths may be structurally truncated. In the past the barrier to mobility between clerical and professional occupations was primarily sex discrimination. Women were raters; men were underwriters. Theoretically, a rater could be prepared by her job and on-the-job training for an underwriting position, although in practice this rarely happened because of the gender-identification of job categories.

The affirmative action victories of the last decade created new opportunities for women to move up. Perhaps a third of the female underwriters and managers interviewed in one compay had entered the firm 10 to 12 years before in clerical positions. Now, however, as lower-level professional functions are automated, the "bridge" jobs between clerical and professional occupations are being eliminated. In the words of one personnel manager, there will be a "quantum jump" between the new machine-linked clerical or paraprofessional categories and the remaining, more highly skilled professional, managerial, or sales work.

Barriers to mobility, then, are not new. But if bridge jobs disappear, a new kind of structural barrier to upward mobility will be created just as the older sexual barrier is being eroded.

A NEW EXPERIENCE OF WORK
AND THE QUESTION OF CONTROL

The new jobs in the insurance industry may not only be dead-end, but also boring, stressful, and deeply unsatisfying. Even fairly skilled work that is driven by the logic of computer algorithms seems to subject workers to unusually high levels of stress (Adler, 1983; Zuboff, 1982; Gregory, 1982). Partially this is the product of what Zuboff called the "curious combination of abstraction and routinization" that characterizes computer-mediated work. Daily activity is centered around the comprehension and manipulation of abstract symbols. Not only is this labor more socially isolating, it requires sustained concentration. Because the fruits of their labor "disappear behind the screen," workers complain that they lose the sense of accomplishing a task (Zuboff, 1982).

Similarly, although labor is being electronically reintegrated, the autonomy of the worker may be diminished. As numerous commentators have argued (see, for example, Noble, 1979; Braverman, 1974; Greenbaum, 1979), the question of control over the labor process brings into sharp relief the extent to which job design is socially determined. The implementation of the new technologies is posing an important set of choices.

On one hand, productive exploitation of the vast amounts of information made available by the integrated data bases would dictate a diffusion of that information throughout the organization. Zuboff (1983) and Hirschhorn (1981) have suggested that this is, indeed, occurring in continuous process manufacturing; and Sabel (1982) makes a similar argument about the effect of CAD/CAM technologies. On the other hand, American managers have an old and deep distrust of their work force, as witnessed by the remarkable proportion of managers and supervisors in U.S. industry.

Today this tension is evident in the insurance industry, both within and between companies. In some situations former data entry clerks are being given wider responsibilities for policy issuance and claims processing and professionals are taking on tasks of data analysis and strategic decision making formerly reserved for management. In other situations task variety has been diminished and the daily work process is closely monitored by machines. For example, in some companies supervisors can keep track of the number of cases handled by each customer service representative and standards have even been set for the number of times the telephone is allowed to ring before it is answered (Appelbaum, 1984).

Significantly, in either case, although the integrated systems may reduce the role of middle management, they provide senior managers with an important new source of control over the daily operations of the organization. And, as Zuboff (1982) has argued, because the integrative logic, decision rules, and operating procedures are designed into the computer programs by a small group of managers, systems analysts, and programmers, users have little comprehension of the system's overall function or the normative criteria on which it was based.

Finally, in this regard, it is interesting to speculate on how a largely female labor force is likely to affect job design in the insurance industry. Although historically one of the defining characteristics of female-typed occupations has been the absence of substantial power over the organization (Oppenheimer, 1968), women have frequently been preferred for highly responsible jobs (such as nursing and secretarial labor) in which authority is extremely constrained. It is possible, then, that in the context of a female work force, lower-level decisions may be pushed down to cheaper labor without threatening top management's control.

CHEAP BUT EDUCATED LABOR

The critical questions when I began this study were not only how is automation likely to affect the lives of the almost 1 million women office workers currently employed by the insurance industry, but also what kind of role might female labor play in the process of economic restructuring. In this section, I would like to offer some admittedly tentative answers to both of these questions.

THE GROWING IMPORTANCE OF FEMALE LABOR

First, despite the fact that female-dominated jobs are disappearing in great numbers, empirically the proportion of women in the insurance industry has grown dramatically in the last decade. By 1982 women comprised 61% of the industry's entire work force, up from 54% in 1970 (U.S. Department of Labor, various years).

Between 1970 and 1980 women claimed approximately 307,000 of the 352,000 new jobs created; and over the next two years women gained 8,900 jobs, whereas men lost 5,500. In total during this period female employment rose by 46% compared to a meager 7% rise in male employment. In contrast to the economy as a whole, where the ratio of women to men increased by 4.7 percentage points, in the insurance industry the ratio climbed by 7.2 percentage points (see Table 7.3).

In this same period, however, as I discussed earlier, clerical workers declined as a percentage of the work force. In other words, the increase in female employment cannot be explained by the disproportionate growth of traditionally female-typed jobs. On the contrary, in fact, what seems to be occurring is a major movement of women into traditional male occupations—professional, managerial, technical, and even clerical.

Between 1970 and 1979 the number of female managers and officers grew from 11% to 24% of the insurance work force; professionals from 17% to 38%; and technicians from 38% to 65% (see Table 7.3). The percentage of women insurance examiners and investigators (formerly male clerical occupations) grew from 9% to 1962 to 26% in 1971 and 58% in 1981 (Appelbaum, 1984).

Overall in the insurance industry, the ratio of women to men in professional and technical occupations rose by 19 and 27 percentage points respectively, as opposed to a 4.3 percentage point gain for women throughout the economy in professional and technical occupations combined. Although the disaggregated comparative

TABLE 7.3
Percentage Point Increase in the Proportion of Female
Employment by Industry and Occupation:
1970-1980

Industry	Occupation				Percentage Total Employment
	Prof	Mgr	Sales	Cler	
Total labor force	4.3	9.5	5.9	6.5	4.7
Manuf	9.0	5.9	11.9	6.7	2.8
Trans & utilities	7.7	9.9	22.8	3.1	3.7
Wholesale	8.8	5.9	4.0	8.9	3.4
Retail	13.4	11.3	5.4	6.3	5.5
Services	0.9	10.1	2.6	11.8	7.7
Fire	13.0	15.7	17.5	5.1	8.3
Insurance*	19.0 P 27.0 T	13.0			7.2

SOURCE: Computed from figures in U.S. Department of Labor (1981 and 1980) and unpublished data from the Insurance Information Institute.
*Data for the insurance industry is from the Insurance Information Institute and is based on EEOC statistics. Although it is not strictly comparable to the BLS data, because insurance carriers tend to be large, the EEOC data on the percentage of female employment is virtually identical to the BLS report. Here, P = professionals; T = technical workers as the EEOC data is captured in this form.

statistics are not extremely reliable, women in the insurance industry during this period seem to have increased their share of professional employment more rapidly than in any other major sector of the economy (see Table 7.3.).

There are a number of plausible explanations for this rapid transition from male to female labor. The first is simply affirmative action victories. Successful affirmative action suits have been waged against insurance companies and, as a result, companies throughout the industry have developed more egalitarian hiring and promotion policies. For example, because of affirmative action litigation between 1977 and 1983 one company studied increased its percentage of female managers from 4% to 33%; professionals from 27% to 46%; and technicians from 29% to 67%.

There is also, however, reason to believe that cheaper female labor is being substituted for more expensive male labor throughout the occupational hierarchy and that women are being used to smooth the process of job redesign and introduction of the new machinery. The professional staffs in the new highly automated personal lines centers of the company where I conducted my intensive case study are so

TABLE 7.4
Percentage of Female Employment in the Insurance Industry
by Industry Segment: 1978

Occupation	Percentage Industry Segment Female Employment			
	All Carriers	Life	Med/Health	Prop/Casualty
Total employment	60.9	53.8	69.2	64.8
Mgr/admin	23.8	20.5	31.4	24.1
Prof	38.0	40.3	42.9	34.5
Tech	65.3	68.2	81.4	50
Cler	92.4	92.0	91.4	93.6

SOURCE: Equal Employment Opportunity Commission (1979).

overwhelmingly female that the chief administrator of one joked that they are under pressure to develop affirmative action goals for men. She explained that the reason they chose to hire women was that women are more "flexible" than men in adjusting to the computer-mediated labor process.

In another company the introduction of computer-assisted underwriting has shifted the bulk of policy processing from a department that is over 60% male to one that is entirely female. In still another the change to a computerized claims process not only brought protests from the older male adjustors, but was accompanied by an increase in the percentage of female employees from approximately 25% of the claims force to over 60%.

Similarly, in line with this hypothesis, the percentage of female professional and technical workers in the industry varies widely by product line. In the more highly automated life and medical/health segments, women's share of employment in these occupational categories is considerably higher. In 1979 40% of all professionals in life carriers and 43% of professionals in medical/health carriers were women, as opposed to 35% in property/casualty companies. Similarly, 68% of all life insurance technicians were female and 81% of all medical/health technicians, whereas women held only 50% of all technical jobs in the property/casualty segment (see Table 7.4).

Significantly, also, although women are moving up in the occupational hierarchy, female wage rates in the industry remain extremely low. In this case study company, for example, female professionals earn only 16% more than male clericals (whereas male professionals earn 50% more); and the majority of white female managers (57%) earn on the average $4 *less* per week than white male clericals. Between 1970 and 1979 in the finance, insurance, and real estate (FIRE) sector as a whole, wages for nonsupervisory personnel fell more dramatically than in any other sector of the economy; whereas total real wages increased by 2%, in the FIRE sector real wages fell by 8%.

In summary, then, drawing together these statistics with the earlier description of changes occurring in the labor process, automation in the insurance industry seems to be creating precisely the kinds of jobs for which women always have been preferred—semiskilled, low paid, and dead end. As a result, the deskilling of professional functions may be acting as a countertendency to the elimination of routine clerical jobs, maintaining employment opportunities for women.

At the same time, as women move into men's jobs the occupational hierarchy is becoming less gender-segregated than in the past. Perhaps, more precisely, as Burris and Wharton (1982) have recently suggested, the more middle-class jobs (managerial, professional, and some technical categories) are desegregating, whereas clerical occupations continue to be solidly and increasingly female. Although this change opens up new opportunities for college-educated women, it probably not only reflects affirmative action gains but also the fact that in the current restructuring process skill levels are remaining fairly high, whereas much of the actual work is routinized and unsatisfying; and in the face of growing competitive pressures, the substitution of women for men across a range of occupations serves to depress wage levels without sacrifices in the quality of labor. In this regard there is a danger, as one personnel manager suggested, that some formerly male occupations may be resegregating female and will be devalued as a result.

FEMALE LABOR AS A LOCATIONAL DETERMINANT

Although a literate but cheap female work force is of growing importance to the industry, feminism and affirmative action gains have made it increasingly difficult to attract educated women into dead-end, low-wage work. White surburban housewives offer companies a partial solution to this dilemma. Because of their household and childcare responsibilities, these women are less career-oriented and, therefore, more willing to accept jobs with limited occupational mobility; they may trade higher wages for flexible or shorter hours and benefits may be less important to them if they and their children are covered by their husbands' plans. These women also, according to clerical organizers, are considerably less likely to be responsive to union initiatives than are minority women in the central cities, many of whom are the sole supporters of their households.[6]

Nelson's (1983) study of the locational determinants of automated office activities (including insurance) concluded that, holding land costs constant, companies have chosen to site these portions of their operations in areas with a disproportionately high percentage of white married suburban housewives. Similarly, in a fairly recent interview an executive of the Fantus Company, a subsidiary of Dun and Bradstreet that specializes in corporate location, argued that automation is raising skill requirements and forcing insurance companies to relocate in order to maintain a low cost but high-quality work force:

> The increased sophistication of word processors and the growing use of computers to handle claims have forced insurance companies to hire more skilled workers than in the past. . . . As demand for this kind of skilled labor grows, recruiting and training will become increasing problems. . . . To be successful [in a market where demand is high], the company must have salaries and working conditions that make it one of the more attractive employers in the area. . . . Because insurance companies tend to have modest pay scales relative to other employers . . . this will create pressure to locate in intermediate and smaller cities [Best's Review, 1979:62-63].

My own limited case work tends to corroborate these observations. Executives in one company explained in great detail the analysis of census data that preceded the siting of their new automated centers; the chosen workers for the first of these to open in a small town in the Northeast were "white housewives of Germanic descent." Similarly, four companies studied had recently moved parts of their operations from major cities to adjacent suburban areas reportedly in search of a higher-quality clerical work force. The company described earlier that is experimenting with homework consciously saw this program as a way to take advantage of the labor pool of educated women with small children. Insofar as skilled labor is increasingly important to the profitability of the insurance industry, then, it is also becoming a central determinant in its location decisions.

OPPORTUNITIES FOR WOMEN WILL VARY BY CLASS AND RACE

For all these reasons, however, although there will continue to be jobs for women in these automated insurance companies, employment opportunities for minority and less well-educated white women may decline sharply.

According to 1979 statistics from the Equal Employment Opportunity Commission, insurance carriers over the last decade have employed a relatively high percentage of minority clericals—22% of their clerical work force as compared to the industrywide average of 17%. Undoubtedly this is because insurance has unusually large numbers of routine, back office jobs, the categories of clerical labor in which minority workers tend to be overrepresented. In 1980, for example, blacks constituted 11.2% of the work force but 21.6% of all file clerks, 17.5% of office machine operators (including 22% of all keypunch operators), and 15.5% of all typists. In contrast, only 5% of all secretaries and 7% of all receptionists were black (U.S. Department of Labor, 1981).

As the data processing centers begin to close and routine clerical categories shrink, minority clericals are in real danger of losing their jobs. Hacker's (1979) study of technological change at AT&T concluded that the single best indicator of a job slated for elimination by automation was the disproportionate presence of minority women.

The threat to minority workers is especially great insofar as the new offices are sited outside the central cities in white suburbs and small towns. For example, the three new personal lines centers described earlier were set up in towns where minorities represented 3.1%, 3.3%, and 14.3% of the population. Finally, also, as Nelson (1982) has argued, the move to teams that involve close working relations among higher and lower level employees may well favor the hiring of "socially compatible" white women.

SUMMARY

Overall, then, although the new technologies may be freeing women from the pink-collar assembly lines and even raising skill levels of the female work force, there is, in the end, little cause for good cheer. For women at the bottom of the clerical hierarchy jobs are simply disappearing. For skilled and particularly white clericals there will be jobs but not opportunities; in this sense, their situation may actually worsen insofar as the last decade had begun to open up possibilities for advancement. Similarly, numbers of college-educated women may make their way into professional and managerial ranks only to find their talents underutilized and undervalued.

At the same time, however, because of the significantly greater opportunities available to college-educated women, the female occupational structure is apt to bifurcate more sharply than in the past, diminishing even further the egalitarian thrust of feminist strategies such as affirmative action. In general, in fact, the new technologies should make it more difficult for women to organize on their own behalf as work is increasingly isolated and footloose.

CONCLUSION

Finally, to conclude, I would like to return to the beginning and place all these observations within the context of the deskilling debate. On one hand, as the postindustrialists predict, a reintegration of the labor process does seem to be occurring in this industry, unskilled labor is being expelled and, as a result, average skill levels are rising. At the same time, however, labor is being degraded, the occupational structure is polarizing, and, although work has been electronically reintegrated, this is not necessarily translating into greater autonomy or task variety on the job; on the contrary, numerous categories of workers may be much more closely supervised.

One explanation for these contradictions perhaps lies in the critical distinction the postindustrialists make between simple mechanization and total automation. Whereas early forms of mechanization relegate the worker to the role of machine appendange, in the advanced stages of automation workers assume responsibility for "controlling the controls" (Hirschorn, 1981). In this account, ultimately, full use of the new productive force—information—depends on its (more democratic) diffusion throughout the organization.

Significantly, it seems unlikely within the next decade that many white-collar activities will lend themselves to the kind of total automation found in continuous process manufacturing plants, even in an intensive paper-processing industry such as insurance. Too many events in the office are unique, unstructured, or (most important) are dependent on interpersonal interaction to permit them to be easily computerized. Past attempts to freeze these activities in computer algorithms often have had negative repurcussions on the effectiveness of an organization. As a result, the postindustrialists might argue, in offices we should expect to continue to see the proliferation of degrading job configurations deriving from partially automated work processes.

Although this is a plausible explanation, the contradictions may be more fundamental in nature. As numerous commentators have suggested, the critical question is not the logic of the new technologies but the element of social choice. Neither the assembly line nor the hierarchical division of labor between men and women were inevitable. In the face of heightened competition and little organized labor resistance, managers may well choose the immediate benefits to their bottom line that deskilling strategies promise. That is, despite the liberating potential of the computer revolution—and probably even ultimately at the expense of full exploitation of its productive capacity—the old and by now well-criticized tendency of U.S. corporations to maximize short-term profits may well prevail.

NOTES

1. The U.S. Department of Labor, Bureau of Labor Statistics (BLS) has given up its attempt to develop productivity figures for the insurance industry in large part because it is almost impossible to adequately measure the inputs from brokerage houses, real estate agencies, and so on.

2. This is a very rough estimate of labor productivity within insurance companies alone; labor inputs from independent agents, brokers, and so on have been excluded. Source: U.S. Department of Labor (n.d., various years), *Economic Report of the President* (1983: Table B-30), *Best's Aggregates and Averages* (1983), *Insurance Information Institute* (1983).

3. For various legal reasons, property-casualty companies were better able than life companies to move their capital into short-term, high-yield investment products.

4. See Note 2.

5. Unfortunately, there is no good time-series data available on employment by occupation in the insurance industry. The 1980 Census of the Population has still not developed an occupation-by-industry table; in addition, occupational categories have been so significantly redefined as to make comparisons with the 1970 Census very problematic. Data is available from different sources for a limited number of years. Occupational data for 1960 are reported by the U.S Department of Labor (1969). These data are comparable to data for 1970 and 1978 reported by the U.S. Department of Labor (1981b). More recent unpublished data is available from the BLS *Occupational Employment Survey of Insurance, 1978 and 1981*. Unfortunately, there are serious and irreconcilable differences between the OES and Industry-Occupation Matrix data. All I can do is report both data sets here. Fortunately, they basically agree on the direction of occupational change and support the expected trends. The OES data, because it is employer-based, may be the most reliable. In any case, it provides the first evidence concerning the crucial later time period when serious restructuring of work in the industry had begun.

6. Based on interviews with organizers for Service Employees International Union #925 and Office and Professional Employees International Union #3.

REFERENCES

ADLER, P. (1983) Rethinking the Skill Requirements of New Technologies: Working Paper. Cambridge, MA: Harvard Business School.

APPELBAUM, E. (1984) "The impact of technology on skill requirements and occupational structure in the insurance industry, 1960-1990." Temple University (unpublished)

BARKER, J. and H. DOWNING (1980) "Word processing and the transformation of the patriarchal relations of control in the office." Capital and Class 10(Spring): 64-99.

BELL, D. (1976) The Coming of Postindustrial Society. New York: Basic Books.

Best's Aggregates and Averages (1983) "Property-casualty edition." Special issue.

Best's Review (1979) "Insurance office locations in the 1980's." August: 62-63.

BLOCK, F., and L. HIRSCHORN (1979) "New productive forces and the contradictions of contemporary capitalism." Theory and Society 7.

BRAVERMAN, H. (1974) Labor and Monopoly Capital. New York: Monthly Review Press.

BURRIS, V. and A. WHARTON (1982) "Sex segregation in the U.S. labor force." Review of Radical Political Economics 14(3): 43-55.

CSE Microelectronics Group (1980) Microelectronics: Capitalist Technology and the Working Class. London: CSE Books.

CUMMINGS, L. (1977) "The rationalization and automation of clerical work." Master's thesis, Brooklyn College.

DAVIES, M. (1975) "Women's place is at the typewriter: the feminization of the clerical labor force," pp. 279-296 in R. C. Edwards et al. (eds.) Labor Market Segmentation. Lexington, MA: D. C. Heath.

Economic Report of the President (1983) "Table B-3." Washington, DC: U.S. Government Printing Office.

Equal Employment Opportunity Commission (1979) Minorities and Women in Private Industry. Washington, DC: author.

FAUNCE, W., E. HARDIN, and E. H. JACOBSON (1962) "Automation and the employee." Annals of the American Academy of Political and Social Science 340.

GLENN, E. N. and R. L. FELDBERG (1983) "Technology and work degradation: effects of office automation on women clerical workers," in J. Rothschild (ed.) Machina Ex Dea. Elmsford, NY: Pergamon.

———(1977) "Degraded and deskilled: the proletarianization of clerical work." Social Problems 25 (October): 52-64.

GREENBAUM, J. M. (1979) In the Name of Efficiency. Philadelphia: Temple University Press.

GREGORY, J. (1982) "The electronic office: a stress factory for women clericals." Prepared for the Office Automation Conference, April.

HACKER, S. L. (1979) "Sex stratification, technology, and organizational change: a longitudinal case study of AT&T." Social Problems.

HIRSCHHORN, L. (1981) "The post-industrial labor process." New Political Science (Fall).

HOOS, I. (1961) Automation in the Office. Washington, DC: Public Affairs Press.

Insurance Information Institute (1983) Life Insurance Fact Book. Washington, DC: author.

MATTEIS, R. J. (1979) "The new back office focuses on customer service." Harvard Business Review (March-April).

MINTZBERG, H. (1972) "The myth of MIS." California Management Review (Fall).

MURPHREE, M. (1982) "Impact of office automation on secretaries and word processing operators." Presented at the International Conference of Office Work and New Technology, Boston.

NELSON, K. (1982) "Labor supply characteristics and trends in the location of routine offices in the San Francisco Bay area." Presented at the 78th Annual Meeting of the Association of American Geographers, San Antonio.

NOBLE, D. F. (1979) "Social choice in machine design: the case of automatically controlled machine tools," in Case Studies in the Labor Process. New York: Monthly Review Press.

———(1977) America By Design: Science, Technology, and the Rise of Corporate Capitalism. New York: Knopf.

NUSSBAUM, K. and J. GREGORY (1980) Race Against Time: Automation of the Office. Cleveland: Working Women Education Fund.

OPPENHEIMER, V. (1968) "The sex labeling of jobs." Industrial Relations 7(3).

SABEL, C. F. (1982) Work and Politics: The Division of Labor in Industry. Cambridge: Cambridge University Press.

SHEPHERD, J. (1971) Automation and Alienation: A Study of Office and Factory Workers. Cambridge, MA: MIT Press.

SIRBU, M. A. (1982) "Understanding the social and economic impacts of office automation." MIT. (unpublished)

STRASSMAN, P. (1980) "The office of the future: information management for the new age." Technology Review (December-January).

TOFFLER, A. (1980) The Third Wave. New York: Bantam Books.

U.S. Bureau of the Census (1981) U.S. Census of Population, 1980. Washington, DC: author.

U.S. Department of Labor, Bureau of Labor Statistics [BLS] (1981a) Employment and Unemployment: A Report on 1980; Special Labor Force Report 244. Washington, DC: author.

———(1981b) The National Industry-Occupation Employment Matrix, 1970, 1978, and Projected 1990, vol. I. Bulletin 2086. Washington, DC: author.

———(1981/1979) Occupational Employment Survey of Insurance. (unpublished)

———(1980) Handbook of Labor Statistics. Bulletin 2070. Washington, DC: author.

———(1969) Tomorrow's Manpower Needs, vol. II. Bulletin 1606. Washington, DC: author.

———(1965) Impact of Office Automation in the Insurance Industry. Bulletin 1468. Washington, DC: author.

———(various years) Employment and Earnings. (March issue) Washington, DC: author.

———(n.d.) Employment and Earnings, United States, 1909-1978. (unpublished)

WERNEKE, D. (1983) Microelectronics and Office Jobs: The Impact of the Chip on Women's Employment. Geneva: International Labour Office.

WEST, J. (1982) "New technology and women's office work," in J. West (ed.) Work, Women, and the Labour Market. London: Routledge & Kegan Paul.

ZISMAN, M. D. (1978) "Office automation: revolution or evolution?" Sloan Management Review (Spring): 1-15.

ZUBOFF, S. (1983) Some Implications of Information Systems Power for the Role of the Middle Manager: Working Paper. Cambridge, MA: Harvard Business School.

———(1982) "Problems of symbolic toil." Dissent (Winter): 51-61.

8

Information Technology
and the New Services Game

LARRY HIRSCHHORN

□ ECONOMIC DATA suggest that the United States has a service economy; but the data don't tell us what is distinctive about services. Is the transition a real one, or simply an artifact of the way in which we categorize economic activity. Daniel Bell (1972) suggested a decade ago that the services create a new experience of work. In the industrial economy, he suggests, people play a "game" against a "fabricated nature" as they struggle to harness a goods-producing technology. In the service economy, by contrast, people play a game with and against other people. In such activities as selling, advising, teaching, and healing people are both the subject and object of the transaction. The subjective problems of another's taste, feelings, and loyalties replace the objective problems of chemical composition, heat loss, and scrap production.

Bell's distinction may at first seem overblown. We have had transportation, education, and banking services for a long time. Indeed, we can argue that the *personal* dimension of such services have declined. Standard services, mediated services (e.g., purchase by phone), and self-service become more important, while the quintessential personal service, domestics, has drastically declined in significance.

Bell's distinction appears forced only because it was premature. The current confluence of the new information technologies with services gives Bell's formulation a new meaning. The new service game emerges as the new technologies bring a competitive and strategic edge to the delivery of old services. Three processes are apparent: the shift from servicing to marketing, the changing boundary between the provider and the consumer, and the more systematic scrutiny of the service provider's skills.

These changes take place, however, in a new corporate context. The same technologies that reshape the service transaction facilitate the growth of local strategic business units within the modern corporation. Such units use integrated information systems to target goods and services to the demographics and economics of particular cities and regions. The strategic service game and the strategic business unit together create a new business planning system in which locale and place become the fundamental categories for marketing and distribution. While production becomes footloose, the locale paradoxically becomes the intense object of business planning. The strategic frame reshapes both service provision and corporate decision making.

In the first section of this chapter I will examine the new service games; in the second, the impact of the new technologies on strategic business units; and in the third, the resulting tensions and contradictions of the service economy that shape urban economic development. The first section is based on case studies of two companies. These case studies come from a larger research project on the impact of office technologies on job design. The second section is based on extrapolations of general social trends, and the third on the conclusions of the first two.

THE NEW SERVICES GAME

FROM SERVICE TO SELLING TO MARKETING

Consider the case of bank X, a major regional bank in the Delaware Valley. It must compete with an increasing number of nonbank institutions for deposits; at the same time the number of financial instruments available for investment and saving is increasing. Senior management realizes that at the organizational "boundary" between the bank and its customer, bank employees cannot simply take in and give out money, but must actively sell financial instruments.

Technological changes are facilitating such a change at the boundary. Banks were among the first businesses to automate paper processing, combining data manipulation with paper sorting. But, increasingly, these old mainframe technologies based on paper-sorting machines and centralized data processing are becoming obsolete. Depositing, withdrawing, and posting are being automated. Similarly, bank customers using automatic teller machines are increasingly able to access the record of their deposits, withdrawals, interest accruals, and account transfer activities. Finally, when debit cards come into general use, paper transactions will fall significantly.

The new technologies have enabled senior managers to redesign the teller's job. Tellers use terminals to access customer records, reducing the chance of fraud and simplifying settlement at the end of the day. Moreover, bank lines are shorter, reducing customer irritation. But as the job becomes simpler in certain dimensions, bank officials are training tellers to answer customer questions about account options and to spot customers who may be potential customers for new bank products. (In a competitor bank tellers are given bonuses for each sale they make.) Bank managers expect tellers to know what financial products are available and which customers are likely targets for a particular product.

The personal dimension of the old transactions between customer and teller has changed. In the past the teller was expected to establish a tone of friendliness and consideration, a climate of support. But now the teller is expected to be aggressive, to see the customer as a potential sales target. The interaction between customer and teller has intensified as a game of targeting and selling replaces the predictable sequences of smooth service. The teller consequently feels less exposed to error along certain dimensions (e.g., unallowed withdrawals), but is more exposed to new ones (e.g., missing a good sales opportunity). Moreover, the new source of error is more subjective than the old one, and is coproduced by the provider and consumer.

Similar changes are affecting the branch bank manager. Because of increasing competition, senior managers of bank X have decided to increase service levels by increasing the numbers and service capacities of branch banks. In the past all branch managers were dependent on the centralized electronic data processing department (EDP) for banking and customer data and could not create data files relevant to their own marketing needs. Frequent tensions between branch managers and the EDP department emerged as the latter provided reports that satisfied everyone's general strategic requirements, but consequently satisfied no one's specific needs. The economies of scale forced it to produce reports geared to the lowest common denominator of interest.

Now, by using minicomputers functioning as smart terminals, branch managers can download certain central files while creating files of their own. (Indeed, at bank X there was much conflict between EDP managers who sought to restrict employee access to microcomputers and branch managers who wished to acquire them.) Branch managers are able to target particular financial products to their actual and potential customers based on an analysis of customer demographics. As branch managers focus their efforts on selling, customers are able to help themselves, using the automatic teller machines for deposit and withdrawal, smart terminals for reviewing their own financial activity records, and, in some branches, touch screen terminals for assessing savings options and the cost of different loans. Increasingly, branch bank personnel will help customers help themselves (e.g., walk them through "what if" questions about loans and other instruments), trouble-shoot problematic systems, sell services, and review branch strategy.

Here, too, the transaction between the customer and the provider has changed. The game between the two is intensified as branch bank managers increasingly market products rather than simply sell and service them. The branch bank manager can no longer succeed simply by establishing a climate of confidence and securing loyal customers. Rather, he or she must aggressively acquire new customers while changing the banking habits of old ones. Indeed, as senior managers give branch managers greater autonomy, they are holding them more accountable for results. The direction of supervision changes: Branch managers have greater freedom to develop site-specific methods for increasing profits, but are now more accountable for profit outcomes.

SELF-SERVICE AND PROFESSIONAL EXPOSURE

Consider brokerage firm Z. Senior management of the firm realizes that competition from banks and discount houses may erode their market position. They need to maximize the time that their brokers spend in sales, rather than in servicing old customers. Consequently, they have introduced an electronic mail system for customer queries. A customer with a question sends an electronic message through a home terminal to his or her broker. The broker reviews all messages at a certain time in the day and writes answers to each customer's electronic "box." Initially, some managers feared that customers would resent such a depersonalized service. But, in contrast, customers reported feeling more comfortable as they now felt they were not bothering the broker by calling at an inopportune time. The brokers, in turn, feel that they are more efficient. They can now bunch customer queries and do necessary research in one time period, and better organize their sales activities. Such self-service activities will be expanded. Thus, in the near future senior management will introduce an entry-order system in which customers will enter their buy and sell decisions on the electronic mail system and brokers will approve and execute the orders.

Senior management has placed some limits on the technological configuration that shapes the electronic message system. Brokers do not work with smart terminals. Management does not want the broker to be able to download customer records and so create a private file of customer investments. They fear that such private records will make it easier for the broker to walk away from the brokerage house with his or her clients. On the other hand, they want to preserve the special relationship that the customer has to the broker, because it is the broker who produces accounts. Thus, for example, when customers have problems with their terminals they do not call the EDP department directly. Rather, they call their broker, who then consults with the EDP department (sometimes in a three-way conversation) to solve the problem.

The new system has produced new relationships as well. Customers can review their accounts' activity and so assess their broker's effectiveness. Missed opportunities and bad decisions immediately become evident and the brokers feel more exposed. Indeed, the broker may feel even more vulnerable to customer scrutiny as senior managers further develop the system. For example, house management has decided to place the house's research bulletin on-line. In the past a customer might ask a broker to assess a particular stock and the broker would simply consult the research staff's current assessment. Yet, in the transaction, the broker appeared to be the expert. When customers can directly access the bulletin, they may feel as competent as the broker. Finally, in response to competition from the discount houses, firm management has just introduced a cut-rate direct access line for customers who do not want to pay commissions for the service of a broker. Although such a service seems to break the special relationship between the broker and the customer, management is confident that the two markets (the commissions market and the discount market) can be effectively segmented. They argue that the reputation of the house is embodied in the broker, and that customers feel safer when they have a broker who will vouchsafe good practice.

But this may prove unrealistic. Professionals may no longer be secure in a service economy as customers become more competent and have the tools to educate themselves. A decade ago many doctors believed that patients would never see a physician's assistant or nurse practitioner instead of a physician. But experience in group practice settings suggests that knowledgeable patients can trust presumably less qualified health professionals for particular services. If the managers of brokerage house Z prove wrong, they may be undermining their own brokerage services, without giving adequate attention to the definition of a new broker role.

As the case vignette indicates, the dynamics of self-service introduce a new competitive and developmental edge into the broker's work. The customer increasingly can scrutinize the broker's performance. The broker no longer controls the information sources and access routes that gave him or her a monopoly of knowledge and supported his or her expertise. Indeed, the house's strategy highlights some of the contradictions in the development of service jobs. While protecting the special relationship between the broker and customer, the house simultaneously undermines it by offering discount services. To protect its monopoly it emphasizes the old special and personal relationship between customer and provider. But to compete in the service economy it facilitates customer self-service and erodes the broker's monopoly control over information. This suggests that in the long run the broker may have to develop a new and perhaps expanded role to retain his or her customers. He or she must enter a game against the customer's growing competence by developing new areas of expertise. He or she may, for example, have to become a financial counsellor helping customers to distribute their investments between a whole range of financial instruments.

IN SUM

The following list highlights some of the critical differences between the old and new services.

FROM	TO
old services	new service
smooth predictable transaction	a game
servicing	selling, marketing
operational focus	strategic focus
loyalty	contingent relationship
dependency creating	interdependence

FROM	TO
standard services	tailored services
expert knowledge	a changing division between expert knowledge and layman knowledge
producer clearly distinct from consumer	co-production of services

As the list indicates, the new services introduce a competitive, strategic, and developmental edge into the transaction between producer and consumer. Self-service based on the new technologies reduces provider expertise, exposes the professional to scrutiny and pushes the provider to develop new competencies and services. In turn, the availability of information enables providers to develop new strategic plans for the development of new services. The tone of the service transaction changes. The old personal tone based on smooth service, loyalty, and dependability gives way to a new one based on a game with and against the customer. The transaction is a contingent one based on competition between the customer and provider, between providers, and between old and new services. There are losses and gains in this change. The customer loses the climate of personal care, but becomes less dependent. The provider loses a monopoly of information and loyal clients, but gains new potential marketing opportunities.

SERVICES AND THE LOCAL ECONOMY: SOME PREDICTIONS

Service provision is becoming a strategic game. Direct providers and their managers increasingly must think in developmental terms as both the new technology and the dynamics of self-service continually reshape their relationships to clients and customers. Thus, as we have seen, at bank X the emergence of the new service game complemented the rise of a more autonomous branch bank manager who developed marketing plans for a particular locale. But this suggests that the senior managers of companies and institutions, who control the strategic planning process and monopolize their access to the necessary data, will decentralize plan making and plan implementation on a significant scale.

Is this plausible? It is, because the decentralization of strategic planning competence is shaped by three other powerful trends: the redistribution of "information capital" in the modern corporation; the breakup of mass markets; and the power of information systems to expose new local opportunities for marketing and profit making. These processes together create local "strategic business units" within the modern corporation. Such units focus their marketing efforts on particular locales and regions and target their services to the particular needs of a local population. The strategic frame of the services game is matched by the planning competence of the local strategic business unit.

This has paradoxical consequences for the role that "space" and "place" play in economic development. It is common to argue that in a service economy, economic development becomes independent of the spatial distribution of activity. Services are footloose. Investment decisions are not shaped by locational economies. We enter the nonplace realm. This argument is oversimplified. It is true, of course, that the new information technologies centralize and concentrate certain services as never before. It matters little, for example, where an information assistance operator is located. Similarly, management elites in both the private and public sectors are developing the tools and competence to plan and manage "megaprojects" on a global scale (e.g., oil pipelines, space industrialization). Yet, at the same time, the local area becomes a focus for strategic planning, as differences and variations between neighborhoods, cities, and regions become the basis for marketing. Services are targeted at local areas and the service mix varies with the variations in local demographic, social, and cultural forces. These developments, in turn, reinforce the strategic game of service provision itself. Let us examine the three trends that are shaping this emerging corporate planning system.

THE REDISTRIBUTION OF INFORMATION CAPITAL

As we have seen, the EDP department at bank X did not want branch managers to buy personal computers without getting its approval. This is a common tension in many companies as EDP or MIS (management information system) divisions lose their monopoly over information and programming resources. Frequently, these central data processing units will convince senior managers to control the purchase of computer hardware in the interests of systems compatibility. They argue that the company will suffer if each unit or division buys its own personal computer so that one division's computer system cannot "talk" to others in the company. Yet even if the company can impose standards on what brand of computer managers can purchase, it becomes increasingly difficult for them to control the rate of computer acquisition. As the price of microcomputers falls, their purchase can be hidden in purchase orders for typewriters and other "innocuous" office equipment.

Finally, recent experience suggests that central data processing units cannot control software purchase and utilization. The quantity of available software is simply expanding too quickly. In the end EDP departments, anxious to sustain the viability and centrality of large-scale computing capability, will have to offer microcomputer-equivalent programs on the mainframe itself (the microcomputer would then be used as a smart terminal rather than a stand-alone computer). Thus, mainframe use will actually be shaped by the software market outside the company. This means that division and unit managers will be able to acquire the data processing and handling programs that best suit their own development and requirements. Central data processing units will most likely lose significant control and function largely as "libraries," testing and supplying relevant software.

Information capital in the modern corporation is rapidly being redistributed. This poses significant dangers to senior management for two reasons. First, increasing numbers of managers can access and manipulate corporate data bases and so propose new strategic plans or critique old ones. Second, the accompanying telecommunications technology (e.g., electronic mail) will make it easier for informal coalitions to coalesce and sustain themselves around alternative strategies and tactics. The corporate planning process may become overpoliticized.

This is unlikely to happen. Instead, senior managers will create many more strategic business units within the corporation through regional, local, and product decentralization. They will resolve the tension between a broad distribution of information capital and a concentrated decision process by giving managers greater local responsibilities. Managers can then use the new information capital to develop local strategic business units, rather than trying to redirect central corporate strategy. Thus, as we saw in this case of bank X, senior management is giving branch managers more autonomy, holding them more accountable for profits on one hand, but supporting them with information capital on the other. This decentralizing tendency is supported by other trends: companies' desire to reduce overhead; senior management's belief that quality and inventiveness improve when managers are responsive to customer complaints and preferences; and, as Chandler (1977) has argued, the emerging diseconomies of centrally managing the divisionalized corporation. The center of the modern corporation increasingly looks like a bank, taking profits from its subsidiaries while distributing investment funds to them.

THE BREAK UP OF MASS MARKETS

As the corporation creates locally based strategic business units, there is a similar break up of industrial mass markets. The latter emerged through two cycles of development: the emergence of the wage-goods industry that supplied the urban working class with food, clothing, shelter; and the development of the single-family home that capitalized nuclear family consumption in the form of cars and appliances.

Four social trends suggest that this latter cycle of development is over. First, the household structure has become more complex as the nuclear family declines in significance; second, suburban development has become more costly, limiting the growth of affordable single-family houses; third, the growth of the two-income family has created a market for consumption outside the home (e.g., restaurants); and, fourth, affluence itself has increased the consumption of services (such as health and education), frequently collectively provided, that improve the quality of life (Hirschhorn, 1979). These trends have led to a growing rate of new product failure as companies adjust their marketing strategies to the new social system.

Economic and marketing theories provide little insight into how consumer needs are created and sustained. Yet it is clear that successful marketing strategies link product images to images of social status and worth. People want something because others they admire have it; this process has not changed. Rather, the culture no longer produces canalizing and simplifying images of self-worth. The image of the independent nuclear family, ensconsed in its private space and linked to the world through the husband and father, provided the matrix for need creation over the last century. But in a postindustrial culture this image is fractured. New conceptions of social life emerge from other parts of globe, lost historical traditions from still others.[1] Cultural life becomes multidimensional, posing new problems for marketing.

The new cycle of development has created a more variegated structure of consumption and production, with markets "breaking into smaller and smaller units, with unique products aimed at defined segments" (Business Week, 1983: 97). Thus, for example, white bread output dropped 15% between 1972 and 1977, while the production of specialty wheat varieties increased 62% (Sabel, 1982). Similarly, the fashions of the 1960s reorganized the men's shirt industry so that the large shirt-making companies had to produce a broader range of styles in shorter runs and with shorter notice (Sabel, 1982). And specialty products account for one-fifth of Dupont's chemical output, while specialty grades in metal and building materials have proliferated. Indeed, a study of the West German economy suggests that the number of new types of products has been rising exponentially since the mid-1970s, while the average demand per type of product has fallen (Bullinger et al., 1981). As the service economy grows, the number of product and service types grows while the variation within product categories increases.

Finally, this break up of mass markets takes place as companies develop flexible manufacturing systems (Hirschhorn, 1984). Such systems can produce short runs of a wide range of products on short notice. A new spoke and wheel system of production is emerging in which the same factory can service multiple regions and locales, each with a distinctive profile of demand.

STRATEGIC INFORMATION SYSTEMS

Strategic planning is linked intimately to the quality and scope of a data base. Business and governmental units produce an extraordinary amount of data to both plan and manage their work. But, to date, few could integrate *already available* information to produce data bases tailor-made for specific planning purposes (e.g., a study of the credit needs of single parents who work in area X). The costs of integration were simply too high. The new technologies of computing and communication are reducing such costs. Over the next 10 years local area networks for cross-computer communication will emerge, office buildings will have central computers and the accompanying wiring for directing communications within and between buildings, and technologists will solve the remaining computer-compatability problems.

However, this new capability will not lead to the growth of more global and comprehensive data bases. When technologists and managers first introduced management information systems in the 1960s, they hoped to create such comprehensive systems so that they could control all facets of company work. This hope proved to be utopian and technocratic. Data bases are organized through categories that can never completely describe behavior and affiliation. The more comprehensive the data base, the greater is the chance that category distortion will create errors of description. This happens through two processes: First, category errors in large systems are compounded through data manipulation; second, it is difficult and expensive to modify large-scale data-based systems. Thus, as the social field changes, the system rapidly becomes obsolete. The system becomes a management misinformation system (Ackoff, 1967). Data bases are useful only when the decision maker develops an anthropological feel for the social field he maps (Graham, 1984). The planning process must be shaped by an interaction between field visits and data manipulation. The decision maker must have direct or indirect access to the local terrain. In this way, tacit knowledge and abstract thinking can be combined. To be sure, senior managers must think globally, but increasingly they are using methods that require less data and better modeling of a limited number of key relationships. To be successful, however, they must stay in touch with their managers in the field. Thus, for example, companies are investing substantial money in their complaint departments. Increasingly, they need direct feedback from their customers about product and service quality; the customer's purchase behavior no longer provides enough information. Thus, paradoxically, the growth in formalized data systems increases the value of subjective knowledge.

This suggests that as managers integrate existing data bases they will produce local data bases tailor-made for specific purposes. The new strategic information systems will be used primarily to *target* the delivery of goods and services, to match the company's offerings to the sociological profile of a particular locale, region, or neighborhood. The strategic business units will be able to develop social and demographic pictures of very small areas (this is the basis for the IRS's new capacity to catch tax avoiders). The new information technologies thus intensify local area planning.

CONVERGENCE

Thus, the three trends converge. Strategic business units within the modern corporation proliferate, their managers use the new communications technologies to map their local targeted markets, while the break up of mass markets rewards the business that can target production and consumption to the demographics and economics of a particular area. Information structures and spatial structures mesh. The new profit opportunities lie in uncovering microvariations between locales and groups, and targeting goods and services for each. The microdynamics of social life are matched marketing strategies. In Marx's terminology, "use values," represented by the social organization of needs, are transformed through information systems into "exchange values." The local arena—the neighborhood, the city, the region—becomes the object of intense planning, targeting, and marketing. The new service game, expressed in its strategic and developmental edge, is matched by the new strategy of local targeting.

Observed developmentally, we can say that in the industrial economy strategy making was centralized, consumption was standardized, and the location of production varied with the "natural" economies of resource availability. In the service economy, by contrast, strategy making is decentralized, consumption is variable, while resource availability only weakly determines the location of production. (Resource economics becomes less significant as the ratio of intermediate services to value added in goods production rises.)

The impact of these three trends is summarized in the following list.

Industrial	Postindustrial
Mass markets	Targeted markets
A focal image of consumption	Many competing images
Centralized strategy making	Decentralized strategy making
Comprehensive data bases	Local, purpose-specific data bases
Formalized information	Subjective or anthropological knowledge
Planning for the average case	Global and local planning interpenetrate
Indirect feedback from consumer to producer through the purchase decision.	Direct feedback via communications system

The service economy is not a nonplace realm. While certain standardized services can be delivered anywhere (e.g., telephone information assistance, credit card operations) places and locales actually become more intense areas for business planning. Politically, this is ironic. Social change theorists hoped that new dimensions of resistance and change might emerge from local and regional struggles against the metropoles, where power and wealth were concentrated.[2] But the modern corporation is also going local, and native. As in the past, urban development and corporate development are inextricably linked. If social struggle is to emerge around the form of collective life and collective consumption in particular locales, it will have to contend with a more intensified and developed business planning system.

IN SUM:
THE DEVELOPMENTAL TENSIONS
OF THE SERVICE ECONOMY

In the new service economy the service transaction itself takes on a new tone and edge; it becomes a strategic and developmental game. The interplay of service and self-service and the shift from servicing to marketing undermine the old service climate of continuity, loyalty, and personability. This new edge is matched, in turn, by the strategic targeting of services and goods to particular locales. This new targeting emerges at the confluence of three trends: the emergence of strategic business units in the corporation, the break up of mass markets, and the development of integrated information systems. The new technologies of information and communication play a central role in this entire process. They facilitate self-service, they create information capital, and they make targeting possible.

Yet these shifts create a new set of developmental tensions and contradictions. Consider the following scenario:

The branch office of a large legal firm is facing increasing competition in the neighborhood it serves. Legal clinics are offering prepaid plans to individuals and groups, while "mediators," who frequently do not have law degrees, help adversaries resolve conflicts without lawyers and courts. They have been most successful in helping people resolve domestic conflicts, landlord-tenant disputes, employer-employee conflicts, and chronic fights between business partners. The emergence of legal plans and mediators seems linked both to the growing complexity of business and community relationships and to people's desire to gain control over how they resolve their disputes. Social workers, underemployed laywers, arbitrators, and a few radical law professors have established an intellectual basis for the mediation profession. In particular, they have developed the concept of preventive negotiation.

The branch responds by giving their clients terminals and access to their programs and data bases. Clients who have relatively simple needs (e.g., composing a real-estate transfer document or a will) can do it themselves, calling on the help of a paralegal when necessary. Clients with more complex needs can search the firm's data base using a structured search process to see which cases in the firm's experience most closely resemble their own. The data base protects the confidentiality of clients by appropriately disguising identifying information. Finally, the firm decides to market this service to mediators and dispute resolution centers so that it can profit from, and perhaps coopt, the very social forces that are changing the relation between the legal system and social disputes. All clients pay a one-time fee to sign up for the service and a charge each time they search the data base.

The corporate office first approves the services, but then orders the firm to suspend it. Corporate managers are worried about three things: First, they believe that the data base may not appropriately seal off information from other branches. There might be information leakage from the corporate data base to the customer data base through poorly specified codes. Second, they believe that the corporation is legally at risk if a client makes a poor decision based on his or her search through the data base. Third, they feel that the branch's strategy dilutes the company's new marketing strategy, which is to highlight the salience of professional assistance in the development of a client's legal plan. They argue that the branch may become too much like an information bureau, with increasingly estranged relations to the legal profession. They believe that the community dispute resolution and mediation movement will be shortlived, as dissatisfied participants ultimately take their cases to court. They have decided, instead, to join an industry association that is lobbying to put restrictions on mediators' practices. Under these conditions they can hardly let a branch offer services to mediators.

Naturally, branch managers protest. They believe that senior management is underestimating the salience of shifts in the social field. Using local government data on divorces and court data on custody and property disputes, they show that there has been a significant drop in the ratio of the former to the latter. People are using the courts to divorce much less frequently. In addition, using an industry association data base, they demonstrate that divorce practice for member firms has fallen. They conclude that mediators have captured the divorce business. Using comparable data bases from other private and public sources, they show that the same trends are operating in the real estate market (people are doing it themselves), landlord-tenant disputes, wills, and incorporations. Finally, they produce information from "current contents" data-bases that suggests that companies are increasingly internalizing their legal costs while trying to mediate disputes and suits with other companies without using the courts. Apparently, industry associations are playing a new role here. They plan to take all this information to the next corporate strategic planning review, but are pessimistic that they can change anyone's mind. They dream of buying out their branch and starting on their own.

This is fiction, to be sure (although there are many signs that a powerful mediation movement is emerging), but it highlights five developmental tensions and contradictions that will shape the development of a service economy. They are as follows: (1) the tension between service and self-service, as competent consumers threaten the status of service providers; (2) the tension between the monopoly position offered by the old personable service system and the developmental and expansionist opportunities offered by the new service game; (3) the tension between centralized control over corporate plan making and the politicization of planning as information capital is more broadly distributed; (4) the tension between the corporate center and the strategic business unit; and (5) the tension between social movements that define new forms of collective life and the more intensified and localized business planning system.

In this senario information technologies shift the balance of power between the consumer and provider, while new conceptions of dispute resolution emerge from the social field. A strategic business unit, anxious to retain its markets, develops a marketing and product strategy that conflicts with central corporate priorities (i.e., consumers cannot be given access to corporate data bases). The resulting conflict can lead in several directions: a political conflict within the company that stalemates its development, the imposition of greater centralized controls on branch offices, a successful industrywide initiative to squelch alternative services, or a strategy of segmenting old clients from the new service system by upgrading old services. This scenario, although a fiction, perhaps can provide guidance for future empirical studies of the service economy.

NOTES

1. This is the central theme of Bell's *The Cultural Contradictions of Capitalism* (1976).
2. Castells has examined this phenomenon in great detail (see Castells, 1983).

REFERENCES

ACKOFF, R. (1967) "Management misinformation systems." Management Science 14 (December).

BELL, D. (1976) The Cultural Contradictions of Capitalism. New York: Basic Books.

———(1972) The Coming of Post-Industrial Society. New York: Basic Books.

BULLINGER, H., H. WARNECKE, and E. HALLER (1981) "Effects of social, technological, and organizational changes on labor design as shown by the example of microelectronics." Presented at the Quality of Life Conference, Toronto.

Business Week (1983) "Marketing: the new priority." November 21: 97.

CASTELLS, M. (1983) The City and the Grass Roots: A Cross-Cultural Theory of Urban Social Movements. Berkeley: University of California Press.

CHANDLER, A. (1977) The Visible Hand. Cambridge, MA: Harvard University Press.

GRAHAM, R.J. (1984) "Anthropology and OR: the place of observation in management science process." Journal of Operational Research 35(6).

HIRSCHHORN, L. (1984) Beyond Mechanization: Work and Technology in a Post-Industrial Age. Cambridge, MA: MIT Press.

———(1979) "Post-industrial life." Futures: 287-298.

SABEL, C. (1982) The Division of Labor: Its Progress Through Politics. Cambridge: Cambridge University Press.

Part IV

The Communications Revolution

9

Communications Technology:
Economic and Spatial Impacts

LIONEL NICOL

☐ PROPONENTS CLAIM that improvements in telecommunications lead to various social and economic benefits: They increase economic productivity and growth and generate new sectors of economic activity; they reduce congestion in large urban centers while facilitating more efficient use of intraurban land; they improve the efficiency of cities while reducing interregional imbalances; and so on.

A review of existing research, however, produces scant proof to support these claims (Nicol: 1983a). No one disputes the fact that communications play a major and growing role in modern economies. Anecdotal evidence and casual observation suggest that advanced economies are fast moving into the so-called "information age." As Webber (1973) observes, "[a] telling mark of the postindustrial economy is the proportionate shift in its inputs from natural to informational resources." This shift has long been under way as shown by the increase in information-based activities in advanced economies—a phenomenon that Porat (1977) dubbed the "information economy." According to Porat's study, total "information activity," including both market and nonmarket transactions, accounted for 46% of the GNP in the United States in 1967. Already in 1967, 46% of the nation's wealth originated in the production, processing, and distribution of information goods and services, conforming to Porat's own definition of the information economy. However, most of the value generated then was in the form of information services produced for internal consumption by governments and firms. Porat's findings thus imply that the production of information services is essentially an ancillary activity. The early growth of the information economy was epitomized by the swelling public and private bureaucracies of the postwar period.

Since the late 1960s the information economy has undergone considerable changes in both size and structure. Today the American service sector employs 7 out of 10 Americans and accounts for 65% of the gross national product. Yet the information economy is on the verge of an even more dramatic transformation. We can confidently predict that the 1980s will witness nothing short of a revolution, perhaps the first serious indication that this coming decade will see the dawning of the information age.

Much has been written about the coming of the information age, but it has yet to be defined and its concrete implications for the economy and spatial structures delineated. I shall emphasize two aspects of this transformation that I contend represent fundamental features of the future information economy. One is the gradual shift of information activities from a nommarket context to a market context. Information services are evolving from ancillary to commodity status. This evolution is clearly reflected in the gradual "unbundling" of the production of information services from the production of non-information goods and services. Perhaps the most distinctive feature of the new information economy is, in fact, the emergence of a highly diversified market for information services, calling in turn for increased specialization in the production of information. Although empirical evidence is lacking, it is not unrealistic to claim that the information services sector will emerge as the largest source of value-added in the United States.

The other fundamental aspect of the transformation pertains to the critical role of communications. Communications has typically been a major impediment to the development of the information economy. Providing communications is often a major cost of doing business. Surveys indicate that communications budgets for large corporations range from 5% to 10% of total revenues and are fast increasing (Department of Commerce, 1983). In a very real sense communications is becoming the most important component of the economic infrastructure. The future of the information economy hinges on building an adequate communications infrastructure. The concept of an integrated communications network is now emerging and, along with it, the idea of planning the communications infrastructure from a local, regional, or national perspective. Communications planning is becoming increasingly relevant as the opportunity cost of nonplanning becomes more evident. Eventually, it will develop into a field comparable in scope to transportation planning today. To date, however, few studies have explored the potentials—and limitations—of the new communications technology. Most questions pertaining to the economic and social effects of communications have focused on one technology: voice communications. Although the plain telephone remains the dominant form of telecommunications service, it is by no means the only one. Furthermore, answers to these questions have been notoriously hard to come by, suggesting perhaps that we are not asking the right questions.

In a world that is increasingly technology-driven, it has become impossible for social scientists to elude the specificity of the new technology. This is particularly true of communications. New propositions concerning the potential economic and spatial effects of communications must be based on a good understanding of the new information technology. The time has come to reappraise some commonly held views of the role of communications in the light of recent technological advances.

I shall first review and evaluate conventional approaches to the effects of telecommunications on economic development and spatial structures, and examine how some of the widely publicized claims of less than a decade ago compare with current evidence. Second, I will examine how information technology is already changing the spatial economy. Finally, I will offer some suggestions for a new approach to communications and its use as a tool for economic development.

COMMUNICATIONS, ECONOMIC DEVELOPMENT, AND SPATIAL STRUCTURES: THE TRADITIONAL VIEW

There is empirical evidence of a relationship between the levels of telecommunications services and economic development. Yet the evidence produced thus far is fragmented, based on highly aggregated data and, most important, essentially correlational. In other words, it offers little explanation of underlying causal relationships. Most evidence is derived from the documentation and estimation of reduced form rather than structural models. For example, a typical procedure is to correlate telephones per 100 population with GDP per capita either for cross-sections of countries or regions at a single point in time or for time series for single countries and regions. All studies show strong correlation between telephone penetration rates and levels of development, but they fail to specify the direction of causation.

The same problem occurs in studies of the impact of telecommunications on location. Geographical proximity has traditionally been advanced as a necessary condition to the specialization of economic activities—the core of urban development. The existence of indivisibilities and the pervasiveness of scale economies in the provision of social overhead capital and the supply of public services, and the need to minimize communication costs and uncertainty in R&D, production, and distribution are often presented as the primary bases for the formation and growth of cities and their clustering into large metropolitan areas. Modern economic development could not have occurred without the increased functional and spatial division of labor made possible by the new technologies. More specialization inevitably means growing interdependence among persons and organizations, inducing, in turn, much larger communications needs. Indeed, it is not by chance that the process of development is characteristically associated with accelerating trends in urban concentration.

Urban concentration alone does not assure economic development. But as long as connectivity requirements are satisfied through the transport of goods or persons (e.g., mail or face-to-face contact), geographical proximity will remain a fundamental determinant of connectivity and, hence, a major factor in the location decisions of firms. Within this context telecommunications—and, for that matter, the telephone—have traditionally been presented as having a decentralizing influence. The basic argument is that a fundamental effect of better communications is to reduce spatial impedance; that is, the frictional forces that geographical space imposes on the transfer of persons, commodities, and information. Developments in transportation technology are largely aimed at "stretching" the ties of proximity over larger areas. Likewise, the telephone is said to have further weakened the need to cluster economic agents and activities.

If all this is true, combined technological advances in transportation and telecommunications ought to lessen gradually the centripetal effect of distance on the location decisions of firms and households. Are these developments heralding the era of fungible geographical space (the day when any point in space would, for all practical purposes, be equivalent to any other point)? This is debatable, but, in my view, unlikely. To the extent that location of economic activities and patterns of settlement are determined by the need to move information rapidly and efficiently, they must be affected by improvements in communications technology. Whether telecommunications will ever bring a unity and uniformity of space is, however, irrelevant. The telephone has probably been as much a factor of concentration as of dispersal of economic activities; it may be operating both ways at the same time. As Gottman (1977) rightly notes, although the telephone may have relaxed locational constraints, "it should be recognized that lofty, dense skylines exist as much owing to the telephone as to the elevator."

Unlike geometrical space, geographical space is not fungible. To be sure, telecommunications have made space more transparent to the flow of information, hence improving the connectivity of places. By changing the mode of representation from printed characters to electric signals and electromagnetic waves, modern communications have virtually eliminated all mechanical links in the transmission of information. More important, perhaps, modern switching technology has endowed information networks with a powerful property: random access. Because of the combinational nature of random access networks, the number of possible two-way interconnections increases as a quadratic function of the number of subscribers in the network. Yet, despite its impressive advantages, there are no tangible signs that telecommunications may be displacing transportation, at least to the extent predicted by the proponents of the wired city (see, for example, Martin, 1978).

The potential for telecommunications-transportation substitution certainly exists and has been extensively documented (see, for example, Harkness, 1973). But the actual occurrence of substitution does not necessarily imply a reduction in aggregate transportation use. Claims to the contrary simply ignore the synergic effects of improved communications on the need for face-to-face contacts that, for institutional or cultural reasons, cannot be handled on-line. The point is that better telecommunications services are likely to both encourage substitution away from transportation and induce new transportation demands. The net effect is an empirical problem. It is fair to say that, at least under present conditions, forecasts of potential telecommunications-travel substitution are excessive. Concerns about whether telecommunications—or, for that matter, transportation—have increased the substitution of places have been misplaced. However, claims that telecommunications have been instrumental in spurring social and economic change are reasonable and, indeed, well-supported by empirical evidence. If these hypotheses are tenable, the potential impacts of information technology on economic development and spatial structure have been vastly underestimated. So important would information and, hence, the role of communications seem for economic and location theory, it is incredible that its study has only received passing interest from social scientists.

COMMUNICATIONS IN ECONOMIC
AND SPATIAL THEORIES

More than two decades ago Stigler (1961) commented, "One should hardly have to tell academicians that information is a valuable resource: knowledge is power. And yet it occupies a slum dwelling in the town of economics. Mostly it is ignored." Except for a few outstanding exceptions, the same still applies today.

The competitive model of economic theory completely bypasses the problem of information by assuming omniscient producers and consumers. Transaction costs—the costs of running the economic system—are often assumed away when, in reality, they represent a major impediment, being in many instances an insuperable obstacle to the formation of markets. Indeed, market failures occur when transaction costs are so high that the existence of the market is no longer worthwhile from an economic standpoint. Were the costs of operating competitive markets actually zero, there would be no economic justification whatsoever for the vertical integration of activities. The fact is that substantial resources are absorbed in gathering information necessary to enter and participate in markets. Information therefore has value and is desirable in itself. Appropriation of that information, however, is costly, partly because individuals have a limited capacity for acquiring and processing information, and partly because access to and diffusion of that information are expensive. The idea behind the social concept of organization is precisely to overcome this bottleneck through specialization in information gathering.

Arrow (1980) has argued that information gathering is probably the most fundamental of all forms of labor division. He also points out an intriguing economic property of information: Unlike conventional inputs, the cost of information is independent of the scale on which it is used. In Arrow's words, "if by joining in an organization a number of individuals can acquire separate signals which can then be used by all of them, net benefits increase more than proportionately to the size of the organization." This statement illustrates the fundamental economic justification for the existence of information networks: the importance of scale economies in information gathering and management. It also highlights the critical organizational function that two-way communications media fulfill within the economic sphere. The new concept goes far beyond the traditional view of the telephone as a purely informational medium, for it adds a productive dimension to communications.

This aspect of communications becomes more meaningful when one considers that the creation of information is an activity characterized by high fixed costs and very low marginal production costs. (A good example of this characteristic is provided by the "packaged" software industry.) This, in turn, has two major implications. First, it makes diffusion—not production—the pivotal component of the cost of information. Second, because of the public-good properties of information, appropriation of the product by its creators is becoming a central issue. As information production and exchange move to the marketplace, the problem of converting information into transferable pieces of property is drawing more attention. Unless the appropriation question is resolved, the economic incentive to create new information content will be partially lost (Dunn, 1982). On both counts, communications technology will play a decisive role.

Early users viewed the telephone essentially as a medium for social interaction, not as a transactional tool. They had little idea of its potential for generating productive activities. Today some 300 billion telephone calls originate from company and service telephones each year in the United States. So dependent has the American economy become on the telephone that most economic activity would probably grind to a halt were a major failure to occur in the telephone system. Unlike the pioneers of the telephone, we are well aware of the economic value of securing fast information exchange. As they did with the telephone, however, we are in the process of exploring possibilities of the new information technology: We are discovering the value of information as a commodity.

If, as Arrow (1980) noted, "the entire economic system can be regarded as a large organization with prices, purchases, and sales as signals communicated," the significance of the new technology for economic theory is substantial. Rapidly converging computer and communications technologies and the subsequent rise of all-digital, interactive networks will deeply affect traditional market dynamics. The trend toward reduction in transaction costs will accelerate as the environment becomes more transparent to information, bringing new markets to life. A more interactive environment also means that economic agents can respond faster to the various stimuli of the marketplace. As information technology pervades all sectors of the economy, market transactions will gradually move into a real time framework. This scenario has several interesting implications, not the least being that of bringing market operations more in line with the competitive model of economic theory.

Theories of location are hardly more relevant for modeling the effects of communications on spatial structure. In the main, location theory defines spatial structure as the outcome of interplay between centripetal and centrifugal forces. Within the intraurban context the location decisions of economic agents are presented as a tradeoff between land and transportation costs. Likewise, efficiency at the intraurban level is defined in terms of a tradeoff between the scale economies associated with urban concentration and the savings in transportation costs associated with increased spatial dispersal. Balancing these costs and benefits yields the "optimal" distribution of activities within the system of cities. Efficient city size is, in turn, defined in terms of a tradeoff between size economies in production and size diseconomies in the form of negative consumption and production externalities. Eventually, the marginal benefits of increased city size are outweighed by the marginal costs, and efficient city size obtains.

In either case, the fundamental basis for economies of scale is geographical proximity and, thus, they are sometimes referred to as agglomeration economies. Because of the overwhelming importance played by physical distance in location models, existing literature focuses almost exclusively on transportation and ignores communications. To be sure, a major rationale for the location of footloose industries has been the minimization of transportation costs. To a large extent, indeed, "the city took the shape dictated by the transportation technologies of the times" (Webber, 1973). However, if we observe current patterns we find a growing number of exceptions to the rule. Conventional approaches are at pains to explain the locational behavior of entire sectors of activity. In many instances the straightforward relationship between distance and costs is lost or considerably lessened; economic activities that are functionally interdependent can operate efficiently despite increasing spatial dispersion. By reducing the friction of geographic space within large metropolitan settlements, improved communication has freed commercial and nonextractive industrial establishments from mandatory location within central business areas. Clearly, other determinants can take precedence in the location decisions of economic agents: availabilty of specialized labor skills, a more favorable or prestigious environment, or simply lower production costs. Nonetheless, the advent of the information-intensive service economies has already shaken up conventional spatial patterns, reflecting the fact that "[u]nlike the natural resource inputs to prior generations of industry, information inputs are easily transported over geographic distances" (Webber, 1973).

To summarize, the contribution of economic and spatial theories has thus far—perhaps for valid reasons—been modest in estimating the role of communications on the economy, its structure, and its spatial organization. I have recommended a conceptual framework for studying the effects of telecommunications on economic development and spatial structures (Nicol, 1983b). Yet the impact of the new communications technology may be vastly more significant than that of conventional telecommunications. In this respect, most studies to date have dealt only with the tip of the iceberg.

NEW COMMUNICATIONS TECHNOLOGY AND THE SPATIAL ECONOMY: EMERGING TRENDS

It is clear that advanced countries are undergoing a major transformation rivaling the Industrial Revolution in terms of change in the organizational structure of their economies. This transformation, epitomized by the shift of national economies from a goods to a service orientation, has already had a profound impact on the organization of their spatial economies as well. Such transformation would not have been possible without the widespread availability of telephone services. Today, advanced economies such as that of the United States are on the verge of another transformation comparable in scope. This time, however, communications not only facilitate but act as the driving force behind the transformation.

Increasing modularity in the way firms conduct operations in the United States demonstrates that higher connectivity across distance leads to a reorganization of functional linkages among economic activities that would not otherwise have been possible. In the past firms were single physical entities with production, distribution, clerical, and management tasks located in the same place. A distinctive feature of the recent evolution of the spatial economy is certainly the development of the multinational firm, associated in particular with the rise of corporate activities. Despite the multiplication of locationally distinct establishments, functionally related corporate activities can function as single productive units. The result has been, in particular, growing spatial dichotomy between high value-added activity centers (e.g., management, R&D) and low value-added ones (e.g, assembly). Bank headquarters may still be in central business districts, but the bank's account management and data processing are performed at remote facilities. Telecommunications alone do not assure spatial dispersal, which has other causes as well. But, just are surely, these developments could not have occurred without the new technology. How is it, then, that literature on the location and relocation of businesses rarely if ever mentions telecommunications—suggesting that connectivity weighs little on intraurban or even interurban business location decisions?[1]

The answer must be found in the multifold nature of communications technology itself. Most studies of the relationship between communications, economic development, and spatial structure implicitly or explicitly refer to one specific technology: voice communication, a technology born in the late nineteenth century when the first analog signal was sent through a pair of copper wires. Except for switching, this technology has remained essentially unchanged since its invention by Bell in 1876. A century later the telephone network is close to achieving ubiquity in the continental United States. There were 230 telephones in use by June 30, 1877, when the telephone was little more than a year old; by July that figure rose to 750, and by the end of August to 1300 (Aronson, 1977). Today there are 189 million telephones in service in the United States, close to 40% of the telephones in use worldwide. The American telephone plant consists of 22,000 switching offices and more than a billion miles of transmission paths—twice the 1970 figures. Nationwide, 98% of the 84 million households and virtually all businesses are connected to this network that includes 6 billion possible connections and reliably handles about 750 million calls daily (Federal Communications Commission, 1981; Department of Commerce, 1983).

Telephone service not only has come close to achieving what Abler (1977) called complete time-space convergence, it has become cheaper in real terms. In fact, we are close to witnessing complete cost-space convergence, because it has become possible as well to bounce a telecommunications signal off a satellite. From the point of view of a satellite in geosynchronous orbit over the United States, every city in the country is, for all practical purposes, the same distance from every other city. In 1920 a station-to-station three-minute call between New York and San Francisco cost $16.50. The same call cost $2.60 in 1960 and $1.30 in 1980 (Abler, 1977). Starting in 1984 AT&T plans to offer an hour of calls anywhere in the continental United States, at night and on weekends, for less than $10.00. Other satellite-based telecommunications networks will be offering telephone service to all major cities in the United States for about .25¢ a minute.

In light of this impressive record, it is understandable that customers and businesses take for granted their access to the telephone. Now voice communication represents but a fraction of the communications needs of businesses. A growing share of business communications today involves the transmission of data. Data communications are growing three times faster than conventional voice services. Revenues from data transmission are expected to double by 1985, and to exceed $10 billion only 10 years after the introduction of the first digital communication facilities such as AT&T's Dataphone Digital Service. The growing emphasis placed on data communications results from the evolution of computer and telecommunications industries. In recent years the dominant orientation of the computer industry has shifted from hardware to software. This trend began when major mainframe manufacturers started to unbundle the software running on their machines in the mid-1960s. It accelerated with the microcomputer revolution, when it became clear to users that software was the driving force behind the central processing unit (CPU). Today microcomputers are designed to meet given software specifications rather than the other way around, an indication that the industry has entered "the software phase": 1983 retail sales of personal computer software in the United States reached $2 billion, double the 1982 level (Department of Commerce, 1983).

Yet there are signs that the computer world might already be moving into the "the communications age." The growing focus on communications is itself the combined outcome of two developments. The first is the trend toward increasing distribution of computing power, as clearly evidenced by the proliferation of CPUs. Once the major impediment to the penetration of microcomputers and terminals into businesses and homes, startup costs have been fast eroding. According to industry statistics, the cost of silicon component has been dropping 15% annually for the past few years, and that of semiconductors 25%. The most important changes, however, are taking place as a result of reductions in the cost of computer memory, which has been tumbling at an annual rate of 40%. Functions that cost $1 to provide in 1979 cost about .25¢ or less in 1984. The introduction of very large-scale integration (VLSI) technology has been a major contributor to this trend. Contrary to early predictions, the concentration of processing power into fewer mammoth central processing units did not take place. On the other hand, the proliferation of small computers by no means signals the end for large mainframes; increased distribution of computing capacity has been accompanied by a concentration of storage capacity.

This second development—the increasing centralization of information—is evidenced by the success of general data bases (e.g.,

Lockheed's Diolog, Dow Jones Retrieval, Mead) now extending to specialized commercial data bases. Because it is important to have a single, continuously updated master file of information (although not necessarily stored in a single location), the economics of information gathering, storage, and retrieval appear to be working in the opposite direction from that of the economics of information processing and creation. These trends are also reflected in the new functional division between stand-alone microcomputers and large "host" computers. Whereas the former have become the primary tool for analyzing and manipulating information and for accessing data banks, the latter are now used more for information storage and management than for data processing. As information sources become increasingly remote from information processing units, the critical importance of communications has become more evident, and is further enhanced by one of the positive advantages of interconnecting the processing units themselves—a technology known as local area networking. Thus, the proliferation of computers has made the transmission of data between spatially dispersed CPUs just as important as its processing. However, because computers generate data in digital form at very high speeds, they have different transmission capacity requirements than the much slower analog voice communications. In fact, connecting two computers through the switched telephone network is akin to connecting two pipelines with a straw.

The American telephone system has thus become the Achilles' heel of the information age, not because of decay, but because new demands are placed on the voice grade lines of the telephone network for which they were not originally designed. Paradoxically, the present situation is largely the consequence of the quality of the original plain telephone system. The telephone plant in the United States is still largely based on copper cables. Voice grade dialed lines can hardly handle transmission speeds over 4800 bits per second (bps), whereas full-motion video signals used for teleconferencing require capacities of 1.5 megabits per second (Mbps). Because digital communications costs, unlike their analog counterparts, are dependent on transmission rates and, hence, on channel capacities (i.e., the larger the bandwidth, the higher the transmission rate, the lower the transmission time, the lower the transmission per million bits), more companies are finding it more efficient to run their own private-line systems. For the first time in the history of the telephone, the number of business telephones declined in 1981, perhaps the first indication that firms were bypassing local loops for their communications needs.[2]

The steady growth of markets for private branch exchanges (235,000 PBXs have been installed in the United States) and local area networks (more than 6,000 LANs are presently in operation) and the booming new communications processing industry provide further evidence that the information economy is trying to circumvent deficiencies in the communications infrastructure. In the chain of information processing activities (i.e., information manipulation and creation/information collection, storage, and retrieval/information transmission and distribution/information use and applications) the weak link is communications.

One consequence of the current bottleneck in data communications is to make geographical space far less transparent to data than to voice. Unless connectivity is improved in terms of both speed (i.e., bandwidth) and access (high-speed digital service is mostly available point-to-point), communications costs will remain the single major impediment to growth of the information services industry. High communications costs, for example, have delayed conversion of the data processing services industry from batch to on-line processing. The introduction of a host of new on-line information services by large, as well as by many smaller, highly innovative companies is likewise hindered by the lack of transparency of the communications infrastructure, despite the rapid penetration of microcomputers and intelligent terminals in businesses and homes. Inadequate communications, in short, is responsible for inhibiting what is possibly the most significant development of the last part of this century: the advent of the on-line economy (Nicol, forthcoming). The on-line economy will become a reality when the most widely traded commodity is information itself, and the marketplace is the information network.

Consequences for the spatial economy are twofold: First, the belief is dispelled that space is fungible with respect to information production and distribution. Although the gap between short- and long-distance costs for leased digital lines has considerably narrowed, on-line data transmission is still very sensitive to distance. More important, most data communications services are provided on a point-to-point basis. We are still a long way from achieving random access in digital communication—a major goal of the integrated services digital network (ISDN). Second, the heterogeneity of communications technologies is emphasized. Data communication itself encompasses very different technologies. Although there is some room for substitution among alternative media, the choice of the appropriate transport technology is largely determined by the nature and purpose of the information being transported. Satellite transmission systems, for example, seem the logical choice for routine communications across and among continents because transmission costs are insensitive to distance within the visibility of the satellite. However, bandwidth limitations, inherent delays, and security considerations make this medium less than ideal for the transmission of sensitive data, or for high-capacity local area networks, the domain of broadband or fiber-optic cables. Whereas satellite circuits seem best for relatively long, lightly loaded links, optical-fiber systems appear best suited for shorter and/or heavier links, with microwave radio's best applications falling somewhere between those of the other two modes (Hardy, 1981).

All trends will converge toward the integrated voice-data digital network before the end of the decade. When in place, the ISDN will have achieved the promise of a truly information-transparent environment. That this dramatic advance will advance the fungibility of space is beyond doubt. Whether, on the other hand, it will actually generate radically different patterns of settlement remains unsure.

THE INFORMATION ECONOMY:
REALITY AND PROSPECTS

Faced with the prospect of a $1 trillion information market by 1990 (Department of Commerce, 1983), and under the growing pressures generated by deregulation following the Consent Decree of 1982, computer and telecommunications equipment and services industries are undergoing a major restructuring. During this fundamental overhaul computer and communications technologies are rapidly converging, inexorably intertwining these once separate industries. The specialized common carriers—GTE Sprint, MCI, Allnet Communications Services, and others recently formed by AT&T—are investing heavily to expand their operations to include lightwave systems, satellite services, digital electronic mail systems, video-conferencing, and other long-distance data communications. Much of the expansion underway at the alternative companies occurs in order to reduce their continuing dependence on AT&T's long-distance lines. Alternative carriers are racing to build their own networks; but building transmission capacity is costly. MCI has already invested $2 billion in its network over the past 13 years and plans to spend an additional $2.5 billion by 1986 to achieve national coverage (Business Week, 1984a). The 5 value-added carriers (Tymnet, GTE Telenet, Graphnet, ITT Domestic Transmission Systems, and Uninet) that offer more efficient transmission services through packet-switched data communications networks are also striving to build their own transmission paths. Value-added carriers have been relying on AT&T's Dataphone Digital Service, which provides most of the transmission links for U.S. packet-switching networks—reaching nearly 100 cities nationwide during 1983 (Department of Commerce, 1983). AT&T will now be providing enhanced services as well through its subsidiary, AT&T Information Systems. The divested Bell Operating Companies (BOCs), now grouped into 7 regional companies, are moving aggressively into digital communication and information services. Approximately 55% of the short-distance metropolitan transmission facilities will be digital in 1984, and this figure is expected to reach 90% by 1990 (Department of Commerce, 1983). At stake is the huge and still growing market for data communications and interactive data applications that foreshadow the on-line economy.

In an attempt to prevent increasing bypassing of its network by private lines or microwave links, Pacific Bell is wiring downtown San Francisco with half-inch fiber-optic cables that can carry seven times as much information as a three and one-half inch conventional copper cable. Anxious to participate in the new deregulated environment, Pactel is also offering to construct a 422-mile, 112-channel hybrid fiber-optic wideband system that would provide cable TV programming and two-way information services to more than 50,000 homes in the Palo Alto, California area. At the same time, the BOCs are pushing for FCC authorization to bill local telephone service according to type of use (i.e., voice versus data communication). Nationwide, 87% of residential customers still pay a flat monthly fee for unlimited local calling. The increasing connection via the use of modems to the local coops of personal computers and data terminals with long holding times and the proliferation of high value-added transaction services with very short holding times, on the other hand, is generating considerable consumer surplusses. As the effects of deregulation set in, the divested BOCs will eventually seek to tap these surplusses.

Computer manufacturers themselves have recognized the importance of communications. Communications become so vital to the computer and, more generally, to the information processing industry, in fact, that it is now a major element in determining which companies will survive (Business Week, 1984b). The computer is evolving from a stand-alone central processing unit to an information system. Business information systems increasingly rely on a combination of computer and communication technologies, and "computer makers are scrambling frantically to gain communications expertise and communications-based products" (Business Week, 1983b). Rather than establishing separate communications divisions, major computer manufacturers increasingly are investing in communications firms. With IBM's acquisition of Rolm Corporation and 60% of Satellite Business Systems, integration has become the trend within the industry.

There are inherent dangers in deregulation and the new competitive environment. One concerns the tendency for owners/operators of distribution media to become the providers of information services as well. As Dunn (1982) rightly notes, "[w]herever the owner of the distribution system also has provided the content, there has been a tendency to exclude independent producers of content." Pacific Bell's Mid-Peninsula Wideband Proposal met resistance because it would have allowed Pacific to provide a range of nonentertainment services—including full-motion video, teleconferencing, computer-to-computer messaging, and high-speed data services—on its own network. The trend toward vertical integration may accelerate with the expansion of worldwide information services such as Videotex and Teletext. Until now the provision of Videotex services involved distinct entities: information and services suppliers (e.g., financial institutions and publishers); systems operators (e.g., telephone companies, CATV systems, newspapers, computer services companies); transmission media suppliers (e.g, telephone operating companies, broadcasters, satellite companies, and cable companies); and terminal manufactures. Growing competition is likely to create strong incentives for the creation of fully integrated Videotex and Teletext systems.

Nowhere are the signs of the ongoing transformation to information technology and the on-line economy more visible than in the banking industry. Fast processing and transmission of information is critical to both domestic and international banking operations. With the growth of multinational corporations, financial centers throughout the world have become increasingly dependent on efficient communication. Multinational, multilocational corporate establishments rely on international banks to provide financial and other information services that accurately reflect changing international conditions. Traditional banking activities now use the most sophisticated telecommunications services available, as evidenced by the creation of dedicated networks such as SWIFT (Society for Worldwide Interbanks Financial Telecommunications) that provide specialized telecommunications services to the international banking community. The stakes can be high. For example, before the possibility of electronic fund transfers existed, settlements were traditionally made at 10 a.m. on the day following the initiation of the transaction. The introduction of the new technology has eliminated the 16-hour float process that used to cost banks over $1 billion annually in lost interest.

The introduction of automated teller machines (ATMs) is dramatically changing domestic consumer banking. Bank of America started with less than 80 ATMs in 1980. It now has more than 900 in operation, substituting them for local branches. This means substantial savings for the bank, if one considers that the cost of an ATM transaction is roughly half that of a human teller (Buyer, 1983). It also implies substantial reductions in staff.

In the process banks have become increasingly aware of the strategic importance of controlling and securing access to transmission paths and distribution systems. The risks of a total dependency on outsiders for their supply of vital communications have led large corporations in recent years to develop their own systems and networks. Bank of America, Citicorp, and Merryl Lynch are building private-line networks, largely for their own communications needs but with the intention of selling raw excess capacity and eventually using it to supply information services on-line.

Deregulation in the financial sector has led to increased dependence on communications as more and more nonbanking institutions begin to compete to provide traditional banking services and banks must diversify their activities to include nontraditional sectors. A logical direction for expansion is in on-line services such as home banking and shopping and data processing, which banks will soon be able to provide through their own communications facilities.

The emerging on-line information services market undoubtedly provides such banks with strong incentive to develop and consolidate their own communications systems. Citicorp has spent $100 million to develop its own fiber-optic network. Merryl Lynch, in joint venture with Western Union, is investing $60 million to build the "Teleport," a project designed to meet the communications needs of companies within the New York City metropolitan area. Bank of America in San Francisco uses its own communications systems to provide home banking services, a field it has pioneered.

The introduction of information technology into the banking sector has not led to dispersal of its activities. On the contrary, the current trend seems to be centralization, along with new functional and spatial divisions of labor. This implies, in particular, the gradual reduction in market-oriented, multiservice branches in favor of more specialized centers that can make efficient use of new communications technologies (e.g., central bookkeeping and data processing centers and regional loan offices). The recent experience of the banking industry foretells the changes that advanced economies will undergo on a large scale within the next decade.

Perhaps the most important conclusion is that, owing to the new information technology, communications are taking on a new dimension and, with it, a new meaning. Hitherto, telecommunications have often been viewed in isolation from the economic infrastructure of a metropolis, a region, or even a country. A new approach is required wherein we think in terms of the more global communications infrastructure. The importance of the new infrastructure is as critical to future economic development as the higway system was a few decades ago to the development of the industrial economy. If, as Dunn (1982) has contended, "[v]iewed as a national resource, the information infrastructure is the most valuable resource a nation has," the potential influence of the communications infrastructure on growth of development and productive activities is awesome. Present conditions point to a rapid emergence of the new information economy—the on-line economy. Communications has been the missing link.

NOTES

1. For example, telecommunications is not even mentioned among the 40 or so items singled out in a recent book as factors affecting firms' locational choices (see Schmenner, 1982).

2. The number of business phones dropped from 48 to 47 million between 1980 and 1981, whereas resident telephones increased from 133 to 135 million during the same period (U.S. Statistical Abstract).

REFERENCES

ABLER, R. (1977) "The telephone and the evolution of the American metropolitan system," in I. de Sola Pool (ed.) The Social Impact of the Telephone. Cambridge, MA: MIT Press.

ARONSON, S. (1977) "Bell's electrical toy: what's the use?" in I. de Sola Pool (ed.) The Social Impact of the Telephone. Cambridge, MA: MIT Press.

ARROW, K.J. (1980) "The economics of information," in L. Destorzos and J. Moses (eds.) The Computer Age: A Twenty Year Review. Cambridge, MA: MIT Press.

Business Week (1984a) February 13.

———(1984b) "Special report." July 16.

BUYER, M. (1983) "Telecommunications and international banking: the political and economic issues." Telecommunications 17(May): 5.

Department of Commerce (1983) U.S. Industrial Outlook, 1984. Washington, DC: Author.

DUNN, D. A. (1982) "Developing information policy." Telecommunications Policy (March): 21-38.

Federal Communications Commission (1981) Statistics of Communications. Washington, DC: Author.

GOTTMAN, J. (1977) "Megalopolis and antipolis: the telephone and the structure of the city," in I. de Sola Pool (ed.) The Social Impact of the Telephone. Cambridge, MA: MIT Press.

HARDY, S. (1981) "Satellites, fiber optics, and microwaves: a cost comparison." Telecommunications 18(6): 47-52.

HARKNESS, R. C. (1978) Telecommunications Substitutes for Travel: Report OT-SP-73-2. Washington, DC: U.S. Department of Commerce.

MARTIN, J. (1978) The Wired City. Englewood Cliffs, NJ: Prentice-Hall.

NICOL, L. (forthcoming) "The online economy: new prospects for growth and economic development." Ph.D. dissertation.

——(1983a) Communications, Economic Development, and Spatial Structure: A Review of Research; Working Paper 404. Berkeley: University of California, Institute of Urban and Regional Development.

——(1983b) Communications, Economic Development, and Spatial Structures: Theoretical Frameworks; Working Paper 405. Berkeley: University of California, Institute of Urban and Regional Development.

PORAT, M. V. (1977) The Information Economy: Definition and Measurement; OT Special Publications 77-12(4). Washington, DC: U.S. Department of Commerce/Office of Telecommunications.

SCHMENNER, R. (1982) Making Business Location Decisions. Englewood Cliffs, NJ: Prentice-Hall.

STIGLER, G. J. (1961) "The economics of information." Journal of Political Economy 69: 213-225.

U.S. Statistical Abstract.

WEBBER, M. M. (1973) Urbanization and Communications: Reprint 90. Berkeley: University of California, Institute of Urban and Regional Development.

10

The New Media

FRANÇOISE SABBAH

☐ THE WORLD OF COMMUNICATION is undergoing dramatic changes. New technologies deeply modify the processes of sending, transmitting, and receiving messages; namely, the communication system itself. Such changes include widespread use of satellites, computers, cable television, pay television, video, electronic mail, teletext, and so on. A galaxy of devices and services transform the capacity of communication and the way of communicating. Technological change is reshaping the mass media. Their audience now has the opportunity of choosing between a diversity of programs and sources different from the traditional TV networks: It can subscribe to a cable TV or to a pay TV; it can play its own videos at any time it wishes; it can also receive, via its roof-top dish, an increasing number of satellite-transmitted programs.

Nevertheless, development of the new media is uneven, depending on their date of entry into the market and/or their financial cost. Currently, only cable TV has overcome the barrier of 30% of market penetration that is the magic threshold that experts consider to make a new medium economically sound (Waters et al., 1981). Home video could reach that level in the near future. In fact, the limit to greater penetration of the market is basically a matter of increasing consumers' purchasing power.

The coming of new communication technologies has been met by diverse reactions. For some observers the current process means the end of the monopoly of media networks, now confronted by an audience increasingly sophisticated and active. Others, along the same line of positive judgment, think that the diversity of media allows greater cultural richness. Yet some critics argue that the new

TABLE 10.1

Sender	Model I Transmitter	Receiver
TV networks	Satellite Microwave	Audience/Screen

Sender	Model II Transmitter	Receiver
TV network Cable TV Video tape/home video Computer DBS STV	Satellite Coaxial cable/optic fiber VTR Phonelines	Audience/Screen Sender (feed-back)*

*In some cases the screen becomes sender in the two-way cable system, such as in the Qube system, so breaking the unilateral flow of mass media, previously characterized by a one-way flow from sender to receiver.

media splits our societies forever between information-rich people, those economically able to have access to the new technologies, and the poor ones, adding to their sorrow a new source of discrimination: the lack of knowledge, that is, of power.

Beyond the schematic views of such ideological debates, this chapter attempts to assess the social, cultural, and spatial impacts of the new media on the basis of a two-fold analysis that considers, sequentially, their technological specificity and their organizational implementation as they redefine a new relationship to the media's audience.

WHAT'S NEW IN THE "NEW MEDIA"?

The most important transformation in the field of media in recent years originates from the fact that the screen of the TV set has become a multifunctional device. Whereas it was a simple receiver of images sent by the TV networks, it has become a support of a variety of messages sent by a diversity of sources. These sources include the televised image, computer data, individually casted movies, telephone-transmitted information, and video viewing.

In other words, Model I has been replaced by Model II, as shown in Table 10.1.

At first sight, everybody has access to all sources of information. At least, everybody able to pay. Provided they have the financial resources required for each medium, most consumers have access to cable TV, pay TV, Television Subscription (STV), Direct Broadcast Satellite (DBS), teletext, computer data, and home video. In fact, most of these new media are increasing their penetration of the market, although at different speeds. VCR is currently the leader in the rate of growth. By mid-1984 10 million U.S households were equipped with at least one VCR, but the figure should reach 15 million within a year, and include one-third of all households by around 1987 (Gelman et al., 1984). Some observers consider video as the most important new mass medium. It is, incidentally, a medium for which production is entirely under the control of Japanese firms such as Sony, Hitachi, and Matsushita (Le Monde, 1983).

Concerning cable TV, 52 million homes are wired to cable (Salmans, 1984), 40 million subscribe to cable in 1985 (The Economist, 1985). Pay TV has also experienced a considerable increase, and in 1981 about 10 million American households subscribed (Moorfoot, 1982). DBS is the most recent arrival on the market, yet it is estimated that about 500,000 homes will be receiving it by 1985, and over 1 million in 1990 (Moorfoot, 1982). STV plans broadcasting six different channels of entertainment, programming, audio, and data by 1986, with a subscription fee of $15-$20 per month. The future for this medium could be even brighter if one included the estimated 15-20 million homes outside urban areas that cable TV is unlikely to reach. Also, costs could be lowered if DBS companies proceed with their plans to switch from high-power to medium-power systems (Business Week, 1984). A sign of the commercial future of DBS is that cable giant HBO plans to enter the field.

Yet, except for the novelty of the technology, is there anything different in the ways of communicating made possible by this new generation of media? In fact, yes. There are four characteristics that make the new media really new.

The first is the flexibility of the system. There are open communication systems that (as shown in Table 10.1) increase their potential and diversity by integrating an ever-growing number of technological inputs. This statement will become more clear if we recall, as an example of the analysis, the process of development of cable TV. As is well known, cable TV was created in order to solve the problems of accessibility of TV signals to areas either too far away or victims of a rugged geography. In 1948, three years after ABC, CBS, NBC, and Dumont Network began operations, some appliance dealers tried to erect antennaes on the cities' highest elevation points, and to transmit signals to different homes by means of coaxial cables. Those homes that were TV-less until then received excellent images through cable: Cable TV (or CATV, for Community Antenna TV) was born as a complementary device to the transmission of the signals coming from the major TV networks' stations (Phillips, 1972).

About the same time, the Federal Communications Commission (FCC) became concerned with problems of signals' interference in the very high frequency (VHF) channels it had assigned to the networks. In 1948 the FCC froze the process of new licensing to offer, in 1952, new licenses on ultra high frequency (UHF) channels. Yet a new obstacle appeared for the development of television (Deluca, 1981): Operating TV sets were unable to receive UHF waves. Thus, although the FCC authorized 1400 UHF stations, only 120 were able to function in 1954. Their development came only when TV makers were compelled to modify the characteristics of their sets to enable them to receive UHF frequencies.

But if cable was, at its beginning, a medium complementary to TV networks, it developed on its own when cable operators began to include in its functioning new forms of transmission of signals; microwaves first, satellite afterwards (Tyler, 1981b). The growing transmission capacity of cable channels and their connection with computer-generated information were also major steps in the coming into full blossom of cable systems.

Yet the legal environment was a decisive factor in the evolution of the new media. Particularly in the case of cable, the FCC's legislation, sensitive to the broadcasters' apprehensions, slowed down for years the development of cable TV. In the commission's own words: "In resolving these issues, our basic objective is to get cable moving so that the public may receive its benefits and to do so without jeopardizing the basic structure of over-the-air television" (FCC, 1972).

Thus, the technical flexibility of cable made possible its development as a complementary device to TV but, on the other hand, the social influence of TV networks severely limited its potential uses when it threatened to be an alternative medium or, at least, a different one in relation to traditional television.

Flexibility becomes a major feature also of the technical evolution of existing media. For instance, the daily operations of TV networks are increasingly dependent on satellites, microwave transmission, and video tapes substituting for films. As tapes do not require processing (unlike films), information delivery is speeded up. Altogether, these new technological means allow the networks to produce an instant televised image. The world can be captured, transmitted, and home delivered any time, anywhere. Only the flexible combination of flexible media, made possible by the new technologies, enables viewers to be fully a part of a universe that they otherwise ignore, except for their consumption of live images.

A second feature of the new media is that the new technologies on which they are based are increasingly cheaper, thus lowering the production costs in relation to the growth of revenue. A good example of such a trend is the evolution of the first established pay TV company, HBO (Home Box Office): Its business really took off only when the earth stations' price was significantly reduced to the point of making satellite-transmitted programs affordable (Tyler, 1981a). Along the same line, lowering computer costs allowed the expansion of data transmission services, and forthcoming low-power systems will decisively stimulate development of DBS.

Third, the combination of cable and computers have ended the predominance of one-way flows. Two-way cable systems allow feedback in the communication network. Yet technology is not enough and the use of feedback systems is still limited, basically aimed, currently, at home security and video-shopping. To date, only the Qube system in Columbus, Ohio is equipped to receive feedback from the audience. A push-button box system is installed in a number of homes to allow viewers to answer "yes" or "no" to the questions appearing on the screen. Yet the Qube system is more a symbol of the new possibilities than a sign of the new directions for the media system.

Finally, the new media are targeted. They do not address an undifferentiated, unknown mass audience anymore. They are aimed at specific households whose tastes and interests are relatively known. The new media's patrons are not "the people" anymore, but differentiated segments of homes. And the effects on identified homes can be computer-controlled. For instance, cable operators or pay TV companies can now scramble their signals to disrupt reception in "bad payers' " homes. The new media individualize their relationships to audience for the best and for the worst.

THE ORGANIZATIONAL STRUCTURE OF THE NEW MEDIA

Technology alone does not produce a unique set of social effects. Its impact is highly dependent on the sociopolitical framework in which a new technology is introduced. Thus, the new media are the result of implementing technological change through a series of financial, legal, and organizational arrangements that must be briefly described before we assess their impact on cities and society.

The most striking remark I want to put forward is that the increasing technological diversity of media source does not bring a parallel broadening of sources of financial and organizational control over the messages. In fact, the new media tend to be concentrated, by and large, in the same corporate groups that already occupied the commanding heights in the communications business. For instance, if we observe the evolution of cable television, the most broadly used of the new media, the following facts provide a clear picture of the interpenetration between cable operators and the major corporations in the communications field:

(1) Cable operators themselves are highly concentrated. P. Kagan estimates that about 50% of the subscriptions market is controlled by 10 corporations (Salmans, 1983).
(2) Very often the cable operators are also the production companies that feed their cable systems.
(3) These cable operators/programmers/producers are closely linked to major financial groups (such as American Express or Chase Manhattan Bank) or to some of the leading communications corporations (Time Inc., Warner Amex, Times Mirror, Westinghouse).
(4) Most cable operators are also connected with the TV networks ABC, NBC, and CBS. ABC (as well as CBS) created cable programming services.

If we examine the list of the 9 largest Cable System Operators shown in Table 10.2 on the basis of available information, we can observe the following:

— Time Inc. owns one cable system (American Television and Communications) and two pay cable services (HBO and Linemax).
— Westinghouse owns Group W Cable, which bought Teleprompter, a cable system. Group W is also related to Satellite News Channel.
— Cox Communications owns Cox Cable. In partnership with Chase Manhattan Bank, Cox is also providing home banking services.
— Warner Communications and American Express have formed one of the major cable operators (Warner Amex Cable Communication), itself linked to two of the main pay TV services, MTV and Nickeloden.
— Times Mirror Co. owns Times Mirror Cable.
— Storer Communications owns Storer Cable.

TABLE 10.2
The Largest Operators of Cable Systems

	Subscribers July 31, 1983
Tele-Communications Inc.	2,297,000
Time Inc.	2,267,000
Westinghouse	1,872,000
Cox Cable	1,379,000
Warner Amex	1,340,000
Storer Communications	1,291,000
Times Mirror	858,000
Rogers/UA Cablesystems	776,000
Newhouse	742,000
Continental	686,000

SOURCE: Cable TV investor

— Vidcom Cable (the eleventh-largest cable system) is a subsidiary of Vidcom International, a programmer company. And Vidcom International shares with Warner and Warner Amex a pay service, "The Movie Channel." Together with Hearst and ABC, Vidcom has formed another channel, "Lifetime."

Business concentration in the new media simply follows the general sequence of business, with established corporations taking control of new technologies when they reach the stage of maturity and profitable development. Furthermore, given the tendency toward deregulation practiced by FCC (MacAvoy, 1977), the "spontaneous" trends of business are now freed from the restrictions they had in the early 1970s: In 1972, the FCC, in its Cable TV Report, forbade participation of TV networks in the cable business. On the other hand, the heavy financial requirements imposed by the FCC on the cable operators to be eligible for the "certificate of compliance," forced many small companies to concentrate to be able to survive, or to sell out to larger groups. This process was the origin of the creation of Multiple System Operation (MSO), a corporation that is the owner of cable systems in different areas. From 1979 on the FCC gradually deregulated cable system operations, as well as the communications field in general, putting into practice the standard "free enterprise ideology," thus trusting the market for the restructuring of communications. For instance, the FCC's Office of Plans and Policy wrote in its 1981 Report: "We believe that in general a policy extending free entry into cable to all interested firms is more likely to contribute to a rapidly growing, technologically dynamic industry that meets consumers' needs."

This philosophy is opposed by other institutions, such as the House Subcommittee on Telecommunications (Consumer Protection and Finance), which considers to be a serious problem the fact that, in spite of the potential diversity offered by cable, the major TV networks still dominate the process of video programming. Tyler (1982) estimates that 40% of U.S. cable systems are controlled by broadcasting corporations.

A new step toward even greater deregulation was taken by the FCC in June of 1984, when it abolished the mandatory allocation of 10% of TV stations' time to new or local events. It also eliminated the limits for advertising time (until then set at 16 minutes per hour) and suppressed the cross-ownership rule (Economist, 1984; Business Week, 1984b). Other measures currently under study would also bring to an end the "Fairness Doctrine" that requires the treatment of important public matters, and the "equal time" rule for political debates. Thus, in spite of some opposition in congress, it is clear that total deregulation and the rule of the free market will be the operating principle adopted by the FCC for the field of communications in the foreseeable future.

Such a tendency, at a time of rapid technological change, will increase the control of corporate business over the media and, through an increasingly sophisticated media system, over people's everyday life. This trend is reinforced by the general lack of awareness and concern at all levels of government for the issue of the new media. For instance, state and local governments have not been able to use their power of licensing to win some influence in the new media system. CATV—which was originally conceived as an alternative medium, closer to people's cultures and to community issues—has generally become an extension of TV networks, both in its content and in its organization. The result of the combination of the FCC's free market ideology and the political shortsightedness of government institutions is the current situation of overwhelming control of the "new media" by the same old interests of the "old media."

THE AUDIENCE OF THE NEW MEDIA

The first change produced by the new media is the need for a solvent audience; one able to pay for the viewing. All new media are pay media, through different formulae, from monthly subscription to payment per service. Generally speaking, the more affluent a home, the more it has access to the new media. Income stratification deepens even further in the multicommunication age.

Another feature is that the audience becomes, at the same time, more selective and more selected. Given the multiplicity of media sources and their growing competition, each medium focuses on some sectors of the audience, the general result being a large variety of programs. For instance, cable systems, taking advantage of their greater channel capacity (some systems can carry up to 100 channels), offer a wide range of programs targeted at very specific audiences: women, children, Hispanics, Blacks, religious programs specialized by religion, music channels for different types of music, art channels, sports, fitness programs, cooking programs, and, of course, movies for different tastes (X-rated movies, American movies, French movies, etc.). So the general trend can be characterized as the narrow-casting of the audience, in a parallel move to what is happening in the radio stations. The audience becomes increasingly individualized and segmented.

Also, the development of home video makes the audience potentially timeless. This is particularly interesting because it happens at the same time that networks' broadcasting emphasizes instant, live communication. The audience of the new media can choose between living second by second in historic time and stopping the clock, desynchronizing real time, to program communication according to its own rhythm.

The new audience could also become, at least at first appearance, an active audience, using the screen to do shopping, banking, or to receive mail or information; and, very soon, to communicate by videophone. Furthermore, as I pointed out in my description of new technologies, it is now possible to activate feedback messages to the sender, as in the Qube system. Yet the use of such systems is limited, and they could be slowed in their development by another deregulatory decision of the FCC that lifts the requirement for cable systems to install 2-way cables.

In sum, the new media determine a segmented, differentiated audience that, although massive in terms of numbers, is no longer a mass audience in terms of the simultaneity and uniformity of the message it receives. The new media are no longer mass media in the traditional sense of sending a limited number of messages to a homogeneous, mass audience. Because of the multiplicity of messages and sources, the audience itself becomes more selective. The targeted audience tends to choose its messages, so deepening its segmentation, enhancing the individualized relationship between sender and receiver. It is this characteristic that allows some observers to proclaim the coming of information freedom with the development of the new media, not only because of the greater capacity for selecting messages, but also because of the active participatory role that new technologies offer to the, until now, silent audience. Yet, as I have stated, technology is framed by its social environment and, therefore, the social effects of the new media and their actual use will depend on a number of other factors, some of which I have already presented in this chapter. It is this fundamental discussion that I will now undertake, advancing some exploratory hypotheses that I hope will stimulate more careful research on these crucial topics, frequently obscured by ideological debates over the future of our civilization.

THE NEW MEDIA, CITIES, AND SOCIETY

The coming of new technologies has triggered a major debate between those who welcome the new communication age and those who fear a stream of dangerous consequences for the social quality of life. For instance, Toffler (1980) believes that new communication technologies will free time for people, who will be able to avoid commuting by working in their electronic cottages. Thus, as people will become sedentary while staying productive, community ties will grow stronger and social exchanges will blossom. On the other hand, critics argue that if information is power, the wealthy will become even more powerful, taking advantage of their financial resources to have access to a multiplicity of communication media. Besides, some radical views consider that the control of large corporations over the new media will enhance political and ideological control of the economic power elite over citizens.

Yet such views are challenged by other experts. The new media's audience is considered to be more active, more independent, because it is able to select between a broader variety of offerings and because it actually participates in the culture of multiple villages, substituting for the global villages of the McLuhan's prophecy (Williams, 1982).

There is, in fact, very tenuous empirical evidence that could provide us with an understanding of the real process under way. Nevertheless, some scattered data sources, including the 1981 Nielsen Report on Television (Nielsen, 1981) allow us to propose some analytical statements.

According to the Nielsen Report, cable TV audiences watch more network television than do cableless homes (Stoller, 1982). Thus, it seems that the variety of sources makes the audience both more selective and more screen-viewing dependent. Using Nielsen's data, Stoller points out three variables that correlate positively with the frequency of television watching: the presence of pay TV in the home, the number of people in the home, and the number of children in the home. While average TV-watching time is 49.5 hours per week, it becomes 58 hours for those having access to pay TV. In fact, TV viewing time has increased in the last six years for all categories of audience, whichever classification we use by age, gender, or type of home. For instance, a home with children had an average of viewing time of 53 hours per week in 1975, and of 60 hours in 1981. On the other hand, teenagers continue to be the group that watches less TV. The Nielsen Report also shows that the greater the choice image sources, the greater the frequency of TV viewing.

Thus, growing diversity of media increases screen dependence. The possibility of choice does not lead people to watch their preferred programs and switch off the rest; it actually increases their addiction to the world of images. As they spend more time in front of the screen, we can assume that they spend less time elsewhere and, therefore, social exchange is reduced. This reduction is likely to take place not only in exchanges outside the home, but also in the private domain. It has been estimated that the time of interpersonal exchange at home for each member of the household is about 14 minutes per day (Waters, 1981). If the multiplicity of the new media increases viewing time, with each household member watching his or her program, it is plausible to see even these rare 14 minutes diminishing. The family reunion around the TV set could soon become an outdated custom.

Furthermore, not only does time for social exchange become residual, there is also a breaking down of the sharing of information and cultural background. As information becomes segmented and individualized, the basis for a common cultural framework is increasingly narrow, making social exchanges more difficult and more rare because of the lack of common codes of exchange.

If we remember that regular working time remains at about the same level and that commuting time is increasing in many metropolitan areas, we will see that people have less and less time to see the real world, and more and more changes to watch it through the screen. The screen will become the reality.

Given the current trends, Toffler's arguments about the potential of electronic cottages to favor community relationships seem to be rather utopian. People are so used to electronic communication that they will find it increasingly difficult to go back to verbal exchange. An audience rewarded every day by the self-gratifying mechanism underlying the media's success is unlikely to accept the reciprocality of rewards inherent to interpersonal exchange.

It is true that the new media offer a much greater cultural diversity. Yet the fact that *all* of these cultural expressions have to be communicated through the screen transforms the culture into spectacle. Our civilization tends to substitute images for experience: Journeys become window viewing and slides become souvenirs; gymnastics a TV program; music a video-tape. The old idea of culture based on a process of symbolic *exchange* does not work any more in the new media. If a radio station can live on broadcasting Beatles' songs 24 hours a day ("The All Beatles Channel"), it is likey that Music Television (MTV) will do the same, with one-style programs (Le Monde, 1983b). To be sure, competition will ensure some diversity in the offering of programs. But given the tight control of a few corporations over the media system, it is likely that the choices will be somewhat narrowed.

Cultural products themselves will be shaped by the characteristics of the new media. Films, for instance, will not be exclusively produced for Movie Theaters' audiences (Business Week, 1984c). Producers know that their profits depend to a large extent on the new media. Thus, the content of film production will be deeply modified. In a similar development, the next generation of musical creation will have to rely heavily on the image and on the image potential of new music to adapt to the rapid expansion of MTV and Music-video. The new singers and musicians will also have to be good performers on the screen. Thus, the new media are likely to condition the entire process of cultural expression.

THEY WILL ALSO AFFECT CITIES

In the first place, there is a close relationship between the location of the audience and its accessibility to the new media. Rural or semirural areas with low-density settlements are unlikely to receive cable or pay TV because of the cost of communication infrastructures to be installed. Also, within metropolitan areas low-income neighborhoods, less attractive for advertising and less able to afford pay services, will be the last to be serviced by the new communication systems. Again, people will be individually and collectively segregated in terms of their income and social status.

Second, the new media will have an influence on the use of urban space. Public space, particularly the most attractive areas of the city, are already very expensive for people to frequent. Housing is expensive in these privileged neighborhoods, parking is rare and costly, and entertainment, restaurants, and stores are economically reserved for those who can afford them. Safety also becomes a rare commodity. Cities are divided between expensive, safe places and cheap, unsafe places. In this particular context, what better choice for people than to consume the city from the comfort of home through the many images provided by the screen. If functional needs can also be fulfilled through the screen (video-shopping, home banking, etc.), there will be no need to move outside the home except for work, school, and some basic shopping. Public space in the city increasingly will be the space for the privileged elite who will have the money, the time, and the cultural resources to enjoy theater, select films, ballet, restaurants, night bars, and so on. Shopping areas will resist the trend; but if people continue to stay at home, in their own world of images and sounds, even of image shopping, stores will have to close their earthly doors and vanish into the new commercial avenues of microwaves or cables.

Nevertheless, the greatest social impact of the new media could be in the field of politics. On this crucial matter there is a raging debate between those who consider that a greater level of available information will stimulate participation and those who fear a growing withdrawal of citizens from their political responsibilities.

It is difficult to evaluate the precise political impact of the new media. We do know that the audience who receives the nationwide messages is shrinking. The new media seems to be responsible for a reduction of the audience of the networks during the prime time hours (Business Week, 1983). And it is during this prime time that the news is broadcast, which is the real ground of national televised messages. Cable News, providing information around-the-clock, is *not* the same kind of message, because it does not mix news and coded messages (including selected images) to the same extent as national news. Besides, the more the new media penetrate the market, the less people read newspapers. Time for social exchanges diminishes. And if we add the specific features of the new media (timeless, self-gratifying, individualizing), we could advance the hypothesis of the new media as a factor favoring political withdrawal. Last, but not least, the concentration of corporate power in the media field is unlikely to diversify political choice in terms of accessibility of a wider range of candidates or options to the media system.

Yet some experts still see the new communication technologies as a tool of reinforcement for democracy, and, more specifically, for participatory democracy, through two-way cable devices of the Qube system type. But such a "black and white" choice could oversimplify the issues (Tyler, 1982). Also, community groups could be able to vote from their neighborhood centers after watching a live City Council debate. Nevertheless, an electronic vote that would not rely on an active grass-roots network of community organizations and citizen participation would be little more than a ritual. And the technology itself forces people to be framed in their choices by the binary reasoning of the electronic medium. Thus, although it is an interesting auxiliary device for enhancing information exchanges, the Qube system alone could not solve the dramatic problems posed to our democracy by the transformation of life into images, and of politics into electronic switching.

NEW MEDIA, NEW AUDIENCE, NEW SOCIETY

The new media have segmented and individualized an audience that is increasingly screen-addicted. Revolutionary technological changes have not freed the audience. The new media have not modified the power structure in the world of communications, they have actually reinforced it.

We still know very little about the effects of the new media. But on the basis of fragmented experience and scattered evidence, we may hypothesize that they will lead to a break down of the private and the public domains, to a reinforced spatial segregation, and to a growing individual seclusion in a world of self-gratifying images.

If, unfortunately, we are right in our social forecasting, then perhaps the latest irony of the new media could come in the form of the revenge of the excluded; that is, of those people who, without having access to the new media because of their low income, will be forced to preserve their social networks and their active political participation while the wealthier will remain passively confined to their electronic cottages. Maybe real-life politics could then prevail over image politics.

Yet this paradox could be equally extreme. In fact, technology and politics, image and life, need each other. The current attempt to escape this necessary interaction could be fatal for both because there is no magical solution to the problems posed by the relationship between science and culture, nor to the tension between our daily lives and our imaginary worlds. The new media are not the cause of our nightmares, nor the source of our dreams. They simply amplify the trends of our society, the devils and angels that inhabit us. Communication technologies do not determine our fate: The new media will be what we decide they will be.

REFERENCES

Business Week (1984a) "Why direct satellite TV is down but not out." August 20: 75.
———(1984b) "Congress slows the FCC's rush to deregulate." August 20: 25.
———(1984c) "Home video: the newest boom in entertainment." August 12: 66.
———(1983) "A showdown at the FCC over television reruns." March 28: 79-80.
DELUCA, S. (1981) Television's Transformation: The Next 25 Years. San Diego: A.S. Barnes.
The Economist (1985) "Wait till next year." January 12: 31-33.
———(1984) "If it's dull, delete it." July 14: 37.
Federal Communications Commission (1972) 36 FCC, 2, 144, 158.
———Office of Plans and Policy (1981) 1981 Report. Washington, DC: U.S. Government Printing Office.
GELMAN, E. et al. (1984) "The video revolution." Newsweek (August 6): 32-36.
MACAVOY, P. [ed.] (1977) Deregulation of Cable Television. Washington, DC: American Enterprise Institute for Public Policy Research.
LE MONDE (1983) "L'explosion video en dossiers et documents' (July).
MOORFOOT, R. (1982) Television in the Eighties. London: British Broadcasting Corporation.
PHILLIPS, M.A. (1972) CATV, A History of Community Antenna Television. Evanston, IL: Northwestern University Press.
SALMONS, S. (1984) "Cable operators take a bruising." New York Times (March 4) Section 3: 1, 22.
———(1983) "Growing debate over cable." New York Times (November 26): 21, 24.
STOLLER, D. (1982) "Growing impact." Cablevision (May 10): 52.
TOFFLER, A. (1980) The Third Wave. New York: William Morrow.
TYLER, R. (1982) "Changing channels." On Cable (February): 13.
———(1981a) "Changing channels." On Cable (December): 15.
———(1981b) "Changing channels." On Cable (November): 29.
WATERS, H. et al. (1981) "Cable TV: coming of age." Newsweek (August 24): 47.
WILLIAMS, F. (1982) The Communications Revolution. Beverly Hills, CA: Sage.

Part V

Theoretical Perspectives

11

Technological Determination and Determinism: Industrial Growth and Location

RICHARD A. WALKER

☐ WITH THE BREAKDOWN of neoclassical hegemony in location theory, interest has shifted from the realm of exchange to the realm of production and, thence, to technology. This has dovetailed with a revival of interest in Schumpeter's ideas about technical innovation as the prime mover of economic growth and business cycles. Meanwhile, technology has come to be viewed by the public as the key to the magic kingdom of regional development and national competitiveness. As substantial technological changes and regional shifts are unquestionably in progress, it is salutory that economic geographers are looking seriously into the subject of technology in the spatial patterning of economic growth (for a review, see Malecki, 1983). This process has only begun. It is not surprising, therefore, that various kinds of technolgical determinism have found their way into the regional debate, such as the notion that high tech industries have a unique locational pattern, that R&D centers are crucial to local growth because of their innovative function, or that the product cycle dooms older industrial regions to imminent stagnation.

I wish to drive a wedge into the cracks in some common ideas about technology and location in order to open up further room for

AUTHOR'S NOTE: *I would like to thank Amy Glasmeier and Michael Storper for commenting on drafts of this chapter.*

debate. I do not do this as an antagonist of the idea of technological determination, pushing a putative "Marxist" line about the monocausal force of social relations. Quite the contrary, I see technology as an essential structuring factor in industrial development and location. Nonetheless, it is necessary to frame the limits of technological determination. Technology must be set against other, equally fundamental aspects of the capitalist economy and capitalist growth, particularly capital-capital (competitive) and capital-labor (class) relations. The collision—or, rather, tension—between the relations and forces of production shapes the course of industrial development. Second, historical outcomes are not mere results of impersonal forces, like the ricocheting of billiard balls. Therefore, determinism must be replaced by a structural-realist view of cause and effect that comprehends the gap between underlying causes and actual outcomes, given the infinite possibility of intervening contingencies, which in history (unlike in a laboratory) never can be controlled. To this must be added the necessary intervention of human consciousness and human agency, of choice and struggle. These render all social history an open system, in which results never may be read off from technology or any other deterministic force, no matter how tight the bonds of social structure may appear (Giddens, 1979; Sayer, 1982a; Walker, 1985).

The first part of this chapter dissects industry cross-sectionally, looking at four technological dimensions of production: product, process, linkage, and division of labor.[1] It lays a groundwork for the rest of the discussion in the specificity of industries along these several dimensions. The second part looks at the patterns of technical change and how they further distinguish the developmental paths of industries and their rhythms of growth. The third section uses these insights about technical structuring of industry and growth to critique deterministic models of the technical imperative in the location of industry and regional development patterns, in terms of the spatial division of labor, spatial linkages and agglomeration, technical change and relocation, and innovation and regional growth.

TECHNOLOGY AND THE ANATOMY OF PRODUCTION

Technology casts industries in different molds and sets them down different paths of development. Nontechnological factors augment these differences. Of course, certain technical and social factors lend commonality to the fates of industries—most obviously, their organization under the rules of a capitalist mode of production. But that does not justify the reductionist view of technique as the outcome of either price ratios (neoclassical theory)[2] or class struggle (labor control theory). The physical character of the technological problem of transforming nature to a usable form has an irreducible effect on the shape of production. I consider four aspects of technology here: product, process, division of labor, and linkages among products and processes.[3]

THE PRODUCT AND ITS USES

Conventional economics focuses on one side of commodities, exchange value. By the trick of marginal substitution all differences of use-value are erased at the moment of consumer choice. But the physical, technical shape of use-values matters. The uses of the bicycle and the banana cannot be interchanged. Neither are the bicycle and the automobile adequate substitutes, nor the trolley and the motorcar. The physical character of each makes for a distinct experience of movement, to which people are not indifferent. Indeed, the introduction of each opened up a whole new range of "needs" among the public (Walker, 1981a). One may apply Rosenberg's (1982) term "learning by using" to this process. Products thus literally embody use-value, rather than simply satisfying preexisting uses dreamed up by the brain of the consumer.[4] Alas, the problem of use-values in consumption has been little theorized (see, however, Lebowitz, 1977/1978; Harvey, 1982; Gintis, 1972).

On the supply side, products as physical, technical entities do not emerge from a void. They evolve from preexisting products and processes, often as the unanticipated results of solving technical problems in the functioning of the product, production of a product, or the effort to meet an unsatisfied need. The transistor was meant to be a better vacuum tube, not the precursor of the microprocessor. The physical properties of products lend a particular thrust to the direction of product evolution, regardless of economic and social forces. "Market demand" is not an adequate explanation for the course of technical development (Rosenberg, 1982: 193-245).

Thanks to the inherent properties of products, markets and product evolution follow different paths for different industries, or what Nelson and Winter (1977) call "natural trajectories."

But technical determination bumps into the social practices of consumption and production. Central to consumption is the relation between the preexisting ideas and practices of consumers and the changing use-values thrown onto the market. Industrialists engage in a war of maneuver with one another and the consuming public over the very definition of wants, in order that their products might be the ones chosen as best satisfying such desires (Ewen, 1976). The "sales effort" is not mere ideological manipulation, however. Consumer practice is shaped through the introduction, use, and promotion of products.

Value and class relations enter into the calculus, as well, in that the cost of products and the distribution of income limit what patterns of use are feasible, whose needs may be filled. Demand must be *effective* demand (Harvey, 1982).

The clearing of product markets rests on a balancing act in which not only value and use-value but use and use-value must be reconciled. The market achieves this end to a remarkable degree, but only through considerable bending of prices, product modification, and forcing of wants; periodically, it fails and the result is glut and forced devaluation of commodity-capital. If products were butter to be squeezed into new shapes at will, much of the problem could be avoided; but they have technical rigidities in consumption and production, the same as capital equipment (Harcourt, 1972). In short, the ordinary product cycle view of product maturation and market saturation (Burns, 1934; Kuznets, 1930; Vernon, 1960, 1966; Hirsch, 1967) must be replaced by a more supple one, less based on analogy with natural aging.

The relation of products to industries must also be reconsidered. Industries are taken as unproblematic in most economic and geographic literature. A first-pass definition of industry is along the lines of discrete outputs. Yet there is no commonly agreed on way of handling joint products and multiple product lines. The Bureau of the Census's Standard Industrial Codes (SICs) are based on rule of thumb and cannot be strictly compared across any one level; in some cases, a three-digit code well defines an industry, in others it is necessary to go to the five- or even six-digit level (Shepherd, 1970: 104). In other words, some product groups cluster into broadly defined industries, while in other cases important differences persist to a high level of disaggregation. Indeed, some of the unevenness of firm size and behavior within industries, which is commonly attributed to organizational strategies, is actually based on the production of slightly different products (e.g., customized versus standardized microprocessors). Worse yet is the problem of technical change in products: When is it sufficient to define a "new" product and/or a "new" industry? This remains an open question on which there is little theoretical guidance.

COMMODITY CIRCULATION, OR INTERINDUSTRY LINKAGES

It will not do to treat all products as destined for final consumption, however. Roughly 60% of industrial outputs are inputs to other industries. Such "productive consumption" raises some distinctive issues. Principally, it has a stronger technological dimension than personal consumption. Inputs enter into production in patterns largely set by techniques in place, and normally there are only a handful of available techniques for arriving at the same product (Gold, 1979; Rosegger, 1979; for reviews, see Hunt and Schwartz, 1972; Harcourt, 1972). A second basis for differentiating industries, then, is by their marked variations in input-output patterns.

At the same time, industries are linked together by commodity circulation, which flows down pathways set, in large part, by technological considerations. An obvious example is the automobile nexus that includes large parts of the oil, steel, rubber, and glass industries. These clusters are of two kinds, component-assembly production systems (autos, computers) and serial processing production systems (oil refining, petrochemical feedstocks, plastic fabricators). Similarly, technical linkages are pathways down which the impulses of innovation travel, although the ultimate connections of any change are difficult to pin down (Gold, 1979).

A special sort of clustering is given by industries sharing similar production techniques, involving a common technological base and common suppliers of machinery. In such cases, innovations in products serving as inputs have greater impact on productivity than process innovations spawned within the industry itself (Rosenberg, 1976: 141-150). A small number of industries producing capital goods using a basic technology may be responsible for most of the technical change throughout a range of industry. The classic case is the metal-working industry's reliance on machine-tool technology and producers (Rosenberg, 1976: 9-31). Another is the early textile industry-machine producers nexus (Hekman, 1980a). Today the impact of electronics technology is being felt across a wide spectrum of industries to which electronic devices are being supplied (de Bresson and Townsend, 1978).[5] One should be alert to how little we yet know about "technology systems"—as indicated by the definitional vagueness of even such astute observers as Freeman et al. (1982).

Despite the force of technology in fixing interindustry linkages, and of linkages on the shape of technical change, they must be set in an economic context. At the simplest this means recognizing that some degree of substitution is possible and that price and profit signals do call forth technical changes over time. At the micro level there is a continual jostling of technical and economic considerations in the strategy of the firm (Rosenberg, 1976, 1982; David, 1975; Gold, 1976, 1979). At the macro level—which is the composite but not the sum of the micro—price structure, technology, and income distribution (rate of surplus value) are interdependent; no one can determine the others (Hunt and Schwartz, 1972; Gold, 1976; Storper, this volume). Indeed, what is significant in terms of the dynamics of capitalism is the ability of the system to mesh these conflicting systems over time in such a way as to avoid sectoral imbalances and the outbreak of crisis (Harvey, 1982).

THE DIVISION OF LABOR

I have so far treated industries as if they consisted of a clear-cut product and self-evident production units. Once the idea of linked production systems is introduced, however, it opens up thorny questions concerning the division of labor. The division of labor must be given its due as a distinct field of technological determination, change, and choice.

All production processes consist of many parts. These may be gathered under one factory roof or split between several workplaces. Technical considerations often dictate which makes the most sense; for example, it makes sense to keep work close together if parts can move from one work station to another by conveyor, or if there is a central power source linked to the machinery by belts. Conversely, it may make sense to separate work stations where electric motors are used or "flexible" automation systems are in place. But there is also an element of indeterminancy and choice involved in the organization of complex production, meaning that it is not possible to determine from technology alone where the detail division of labor (within the work unit) will leave off and the social division of labor (between work units) will begin, and which will be embraced by a single "industry" or firm. Our first- and second-cut definitions of industries will not suffice, therefore, because "products" and "linkages" vary by where one draws the line around production. Textile machinery making was part of cloth production until it spun off as a separate industry; and locomotives were part of the latter until they, too, split off. Should steel be produced as part of automobile production, as was done by Ford at River Rouge, or as a separate industry? Should parts move by hand truck or interstate trucking? It will not do to say that the market will decide which is cheapest; what determines cost structures? And it is not merely a cost issue, because labor militancy often provokes capitalists to divide and scatter their workforces, and even to operate redundant plants. Volumes have been written on industrial organization, but we still do not have a theory about why integration proceeds readily in certain industries, certain firms, but not in others (Caves, 1980; Scherer, 1970: chap. 4).

As Rosenberg (1982: 76) notes, "it might be [that] technology flows have radically reshaped industrial boundary lines, and that we still talk of 'interindustry' flows because we are working with an outmoded concept of an industry." Hence, it is better to speak of "production systems" in many cases, as noted above. But it can matter substantially how circulation is organized: A production system embraced by a single corporation is different from several firms linked by the market; a subcontracting system, involving both big firm domination and a special kind of marketing arrangement, is different yet (Murray, 1983). Allen Scott calls this the problem of "integration-disintegration" and focuses attention on economies of "scope"—as distinguished from economies of scale—in production (Scott, 1984; compare NRPB, 1943). But the issue is far from adequately understood.[6]

The problems introduced by the division of labor become even more complex when we include the "indirect labor" components of production systems, such as process engineering, product development, repair, and management (Walker, 1985b). Some technological structuring is discernible, such as the need for long-lived machinery or consumer durables to be repaired or the role of R&D in industries with rapid product proliferation, such as microelectronics. But the economic side of choices about, say, whether to include accounting within U.S. Steel or contract it out, or to commit resources to product differentiation and advertising rather than real product innovation through research, is substantial. A fourth-cut definition of industry might turn on the amount of labor devoted to activities and occupations of an indirect nature across product sectors; this is the strategy chosen by Glasmeier (this volume) to define high tech industry. But it, too, is beset with ambiguities, such as whether technical workers are involved in product development, process change, or marketing. A rather different classification, which cuts across commodity sectors, has arisen to deal with the rise of indirect labor in distinct workplaces, as when one speaks of "the office industry."

THE PRODUCTION PROCESS

Having surveyed some ground often passed over in treatments of technology in industry, we can now take up the production process inside each workplace. All production involves human labor and, as Marx (1967) argues, every labor process has two sides: the production of use-values and the generation of value and surplus-value. The latter is the raison d'etre of capitalism and production would not be undertaken without it. Marx makes a powerful case for the devastating effect of the search for surplus-value on the worker, as well as for the way it drives capitalists to raise the productivity of labor.

Marx's analysis of the labor process still stands as the best treatment of the subject. He does not make the mistake of equating technology with machines, although he understands that "modern industry" revolves around "machinofacture." Instead, he treats three aspects of the application of labor: cooperation, division of labor, and mechanization. Technical change centers on the worker creatively wielding a tool to transform materials into a useful product. They key to the industrial revolution, therefore, is not steam power, but successfully capturing the unique capabilities of human activity—the hand, the eye, the creative mind—in the workings of the machinery.[7] The preconditions for this achievement are cooperation (bringing together many workers) and the detail division of labor (breaking down complex processes into simpler ones and rationalizing them according to mechanical principles). As Freeman et al. (1982: 70) observe, the reorganization of production is as important as the application of technique. The results of this process are, however, the reduction of the former craftworker to a detail worker and ultimately a machine-tender. The process of deskilling has been much in vogue since the effort of Braverman to apply Marx's analysis to the twentieth century.[8]

Nonetheless, Marx's treatment—or, at least, the common understanding of it—needs to be amended in several ways. First, while the tendency to deskill labor is a profound one that runs across all labor processes, the effects of division of labor and mechanization are not uniformly adverse to technical skill: They may lead to reskilling through the command of more specialized tasks (including whole new branches of labor, such as R&D) or the oversight of whole machine systems (Storper, 1982).[9] And, of course, new products and labor processes are constantly being created as the old evolve.

Second, one cannot look at production only as a labor process. It is also a materials-transformation process. The failure to distinguish

between the two aspects of the matter has led to no end of confusion.
As a labor process, production has common features that span all
industries, all workplaces. As the transformation of specific materials
into specific products, within the bounds of physical laws, it is
characterized chiefly by uniqueness and diversity. So are the concrete
tasks and skills of the workers in each industry; as Sahal (1981: 59)
observes, "technical know-how is largely product and plant specific"
and is not readily transferable to other industries or places.[10] The
distinction between materials-transformation and labor process
accounts for the gap in the literature between the "choice of
techniques" debate (for reviews, see David, 1975; Uselding, 1974;
Kennedy and Thirlwall, 1972) and "labor process" studies (for review
see Elger, 1979). If we assume that the choice is a simple one between
more and less capital-intensive methods, and that these represent
movements along the vector of mechanization, then the two sides
collapse to one. On the contrary, there are several vectors of
mechanization (or greater input productivity) possible for an industry.
The choice of technique involves jumps both between vectors and
along vectors that are themselves discontinuous.[11] In any case, the
neoclassical vision of smooth substitution of factors along a tidy
production function does not hold (Rosenberg, 1976: 61-84; Hunt and
Schwartz, 1972; Harcourt, 1972; Gold, 1979).

The material differences among production processes, like those
among products, set industries down very different paths of
development. Innovations in processing, handling, and monitoring
have a materials component that has a logic of its own (as, for
example, how to make a part stronger, more malleable, or accepting of
more electrical circuits). Technical considerations also establish the
possibilities and set the limits on the course of division and
mechanization of the labor process. The tendency to revolutionize
production runs into the concrete technological diversity of the
products to be made; the problems inherent in automating garment
and petrochemical production are worlds apart. Some industries
never advance much beyond the craft stage: Ships, one of the oldest
products, are still made by small-batch process; scientific instruments
may be "high tech," but their production still is labor-intensive (Oakey,
1983).[12] Finally, material considerations may impel as well as impede
mechanization: For example, quality control in microchip production
demands greater mechanization (it might well require more labor in
another industry, however). One cannot explain the radical differences
among industries in growth of productivity (Kendrick, 1973) except by

reference to different technical potentials inherent in specific products and material processes (Nelson and Winter, 1977; Sahal, 1981).

The disjunction between the two sides of production also addresses the riddle of bias in process change (see, again, David, 1975; Uselding, 1974; Kennedy & Thirlwall, 1972). From the materials side, technical change will be neutral as there is no incentive to save on one input more than another (Salter, 1966). From the labor side, however, mechanization always involves labor-saving (greater output to labor ratio).[13]

A third consideration is to recognize the irreducible human element in all labor processes, regardless of technological proficiency and the drive for surplus-value. While people may be constructed along common lines, lending a homogeneity to all labor processes, labor can never be entirely rationalized and given over to machines. The creative human element remains, if only in seeing that the machines perform as programmed (Aronowitz, 1978; Cressy & MacInnes, 1980; Manwaring & Wood, 1984). Every labor process demands special skills that can only be learned through practice on the job. It also requires that workers mobilize their labor and exercise their creativity. Production requires not merely choosing the right mix of "labor inputs" to fit technically-given tasks, but hiring and molding, through the experience of work and managerial control systems, the kind of labor force that will get the job done and done well. This is no mean trick. It only happens through a process of class-structured maneuver and conflict that includes group socialization, the application of managerial power, and various forms of worker resistance within the context of the job to be done, the leverage given workers by their skills, labor market conditions, product market conditions, and the like (Storper & Walker, 1984). Because there is no unique outcome to the social order of the workplace, labor demand cannot be read off from production technique and organization. The concepts of strictly technically determined "skill" levels and "marginal products" of labor are untenable.

The variable condition of the employment relation augments the differences among industries. Furthermore, as has been frequently observed, the state of labor relations affects the course of technical change. Machinery has often been used to break the hold of skilled and/or unionized workers; conversely, militant workers may prevent the introduction of new techniques or docile and low-paid workers blunt the capitalist's drive for technical change (Marx, 1967; Rosenberg, 1976: 117-120). Thus, the path down which technology travels and the rate of movement depend on the social relations of employment.[14] The intervention of class conflict and maneuver into process of technical change add a further element of uncertainty and disjointedness to the flow of innovation, beyond that which comes from purely technical considerations of the imperfect meshing of technology and market.

Fourth, the course of process evolution, like that of product development and the division of labor, is affected by (output) market conditions. A product may lend itself physically to mass machine production, but will not be so made if no market exists. For example, U.S. semiconductor manufacturers have long held that shifting market conditions, owing to rapid product innovation, do not justify heavy investment in mechanized chip production. On the other hand, falling unit price (value) may create a mass market where none existed before, as has also happened in semiconductors. In fact, supply side changes in technique (and other variables) are constantly altering the cost structure of production, altering the shape of effective demand. In short, demand does not just call forth supply; supply can also generate its own demand. The dynamic interaction between the two is the place to focus attention, instead of the fruitless quest for linear cause and effect from one putatively independent variable to another (Rosenberg, 1982: 231-232).

Finally, I should touch once more on the troublesome problem of defining industries. There is a wide range of technologies in use in every industry. Some of the differences are due to mixing of different products (overaggregation), some to the age of fixed capital (a vintage problem along similar lines of development), some to the complex division of labor in most sectors (mixing direct and indirect labor processes). Nonetheless, firms within a single industry are able to set themselves on different courses, some of which alter the course of the industry. Averaging across such differences as if the development path of an industry were a certainty, along which there are leaders and laggards, hides the role of human agency in the process of technical change.

In sum, a multitude of possibilities emerge from the interaction among technological (product, process, and division of labor), use-value, value, and class relations. Therein lies the wellspring of divergence between industries—and of the problems of industry definition. Despite the technical sources of industry differences, therefore, it will not do to call their paths of development "natural trajectories," for the social element looms so large in the history of every industry, from worker resistance to consumer acceptance. It is worse yet to jump from natural trajectories to organic analogies of growth, as in the product cycle. The next section takes up the question of the technological impetus to industrial development and the effect of technology on the pattern of industrial growth over time.

TECHNOLOGY AND INDUSTRIAL DEVELOPMENT OVER TIME

THE PROCESS OF TECHNICAL INNOVATION

We must first dispose of the idea that industrial growth is triggered by major technical innovations—the big bang theory. One finds this idea throughout the literature on innovation diffusion and the product cycle. It rests on fundamental misconceptions as to the way technological change proceeds and how it fits into the wider regime of accumulation.

Some principles of technical change can be enumerated that belie the big bang theory, even while strongly supporting the idea of technological determination in the course of technical change. That is, technical change is, to a significant degree, internally generated by work to solve technical problems and internally structured by the nature of the physical products and processes of production.[15]

(1) Technology does not flow from science,[16] as in the linear view of causality embodied in the trilogy "invention-innovation-diffusion." Philosophically, the trilogy rests on an idealized view of knowledge as the product of contemplation rather than practice (labor); ideas appear in the head of the theoretician and are then applied by the manual worker. This is not even true of science, which requires quite a bit of hard work, let alone of knowledge in general (Bhaskar, 1978). Sociologically, it involves a fetishism of science as true knowledge and scientists as priests who function in a different way from ordinary mortals (Sayer, 1981). Industrially, it simply does not square with the evidence.

The history of industrial technology, from the steam engine to the modern airplane, is that it usually runs ahead of scientific understanding of the underlying principles involved (Rosenberg, 1982: 141-159; Sahal, 1981: 30-32). It is not unusual for industrial engineers to develop a solution to a problem for which there is no scientific explanation, and for that to spur scientific research (Rosenberg, 1982: 126). Industrial technology also gives to science most of its instruments of investigation, which are critical to posing questions as well as verifying solutions. Conversely, the performance characteristics of industrial technologies cannot be predicted well by theoretical science (Rosenberg, 1982: 122). This is to be expected;

according to the realist conception, scientific explanation deals chiefly with underlying mechanisms abstracted from intervening contingent causes, and not the complexities of real situations in which the latter are heavily implicated (Bhaskar, 1978; Sayer, 1982a). It is one thing to know how gravity affects the fall of the apple, another to design an apple picker.

Technical innovation derives principally from practical experience with production. Practical problems are encountered, practical solutions are proposed. "Innovation [is] a process of learning by experience." Moreover, "the process of learning tends to be technology specific" (Sahal, 1981: 37). Few operating production processes are based on formal blueprints, especially after long periods of adaptation (see below). As one plant engineer complained: "If we waited until the designs were completed, we would never start building" (quoted in Piore, 1968: 605).

This does not mean that innovation only takes place on the shop floor, that research and design workers may not be a separate group of workers laboring in another corner or even another building, or that industrial engineers do not refer to scientists and scientific journals as they search for solutions (Price and Bass, 1969). Nonetheless, "the gist of science lies in indicating what is *not* possible" (Sahal, 1981: 62). Practical inventions do not, as a rule, flow from scientists and little industrial R&D is basic research. Industrial engineers have to interact with the production line and the marketing office, and good ideas come from ordinary workers all the time (Piore, 1968).[17] Working production processes are rarely captured in formal blueprints (Piore, 1968).

One must also carefully consider the conditions under which scientific principles may be applied to production—e.g., after a degree of rationalization and division of labor (Marx, 1967; Rosenberg, 1982: 39-51). Despite all the attention to R&D in the literature, there is little evidence that R&D effort explains technical change (for reviews, see Mansfield, 1972; Kennedy and Thirlwall, 1972, 44-50; Sherer, 1970). This is partly because so much invention is happenstance and has unanticipated consequences (Jewkes et al., 1959; Peters, 1983). But it is mostly because R&D and rate of innovation are not independent variables; they depend on the nature of the technology: "The inventive performance of an industry is determined mainly by the nature of its technology" (Sahal, 1981: 57; Phillips, 1971). Promising industries have both high levels of R&D and high rates of technical change; hence, R&D is highly concentrated in a handful of industries (Mansfield, 1972).

(2) "Inventions," do not burst, fully formed, upon the industrial stage. Technical change is made up of a stream of small innovations and incremental improvements in products and equipment, not a handful of revolutionary ones. This is to be expected from a process based on practice and tinkering. Technical change is part of a larger process of improving production through experience, which has been dubbed "learning by doing" (David, 1975; and literature cited there). Even the best idea needs a long series of improvements to perform well. Virtually all inventions require years to be adopted initially (a commonly cited figure is 11 years),[18] years more to be adapted widely, and longer to hit their peak performance. All of these lags result from the need to improve or alter a basic product or process for different specific uses (Gold, 1979; Rosenberg, 1976: 71-73). As Gold (1976: 2) observes, "innovations trigger a continuing process of changes. . . . It is through the ensuing complex of interactions that the innovation is 'digested' by means of progressively more far-reaching adaptations and its effects thereby diffused through the system and over time . . . as a result, the distinctive effects of the given innovation become increasingly indistinguishable." Because of the time lag, several innovations are likely to be in the process of digestion at any time. The reach of an innovation will depend on its adaptability to related activities; apparent pathways may be blocked. Both major and minor innovations are clustered unevenly (Sahal, 1981: 59). Such events are unpredictable because of the inevitable dialectic of dull labor and creative breakthroughs, of human effort and the possibilities inherent in a technology (Sahal, 1981: 41; Jewkes et al., 1959; Peters, 1983).

Because learning and innovation are object- and process-specific and build on one another, technical change is "localized"; every technological choice circumscribes the course of further development (David, 1975: 55-91; Sahal, 1981: 199). This adds further force to the idea of restricted industry development paths.

(3) Technology creates its own "compulsive sequences" of problems and solutions (Rosenberg, 1976: 112). This pushes industries even harder down technically structured paths. The basis of such sequences are the technical complementarities between related parts of a single machine, of a unified production process within a factory, or between different sectors of the social division of labor (Marx, 1967; Rosenberg, 1976: 110-117, 201-206; 1982, 56-62, 70-80; Nelson and Winter, 1977; Piore, 1968). An improvement in one area reveals the inadequacies of another, as when miniturization so reduces computers that the conventional TV picture tube becomes a barrier to portability. Or the imbalance may be felt in terms of bottlenecks in the

flow of materials from one process (or industry) to another. There is no reason to expect that such imbalances can ever be eliminated, moreover, correction in a lagging sphere may catapult it ahead of the formerly leading sphere.[19]

Another kind of complementary technical change is that between substitute products or processes. Not only do new techniques not burst full born on the market, old ones often undergo dramatic improvement under the spur of competition or benefit in other ways from the environment created by the new (Rosenberg, 1976: 202-206; Uselding, 1974: 186-188; Mak and Walton, 1978).

Finally, there are pathways of diffusion between linked technologies and industries, by which technical change in one area helps improve a related process or product (see above).[20] Because of its experiential base, however, technology is not a universal bag of tools that can always be transferred easily from one industry or country to another. Diffusion follows pathways of use; it requires improvement and adaptations; and it does not follow an S-curve of growth (Sahal, 1981: 106). By the same token, followers often do better than leaders in innovation because they are not trapped in an environment and experience that restricts a technology of widespread significance to a truncated use.

(4) Despite the incremental nature of technical change, the sum of such changes is not incremental: That is, there are basic design frameworks within which work advances. Oft-cited examples are the steamboat, the tractor, and the DC-3 aircraft. Sahal (1981: 64) calls these frameworks "technological guideposts," Nelson and Winter call them "technological regimes": What they are advocating is a "structural" view of technological systems, in which an underlying pattern lends coherence to many small changes. "Evolutionary changes do not just pile up. They inevitably build up into a system" (Sahal, 1981: 64). This similar to recent theories of how science evolves (Kuhn, 1970; Piaget, 1970; Bhaskar, 1978). It gives further meaning to the idea of technical determination of industry development paths.

Because technical progress is structured by the nature of the product and process, the possibilities for improvement in a basic design can be exhausted in time. There is, therefore, a limited range within which something like the product cycle can take place. Recall, however, that the level of process development (especially mechanization) to which production "matures" depends on the kind of product and market.[21] Nonetheless, technological exhaustion must be balanced against technical breakthroughs.

(5) Given the structured nature of technological change, it is possible to speak of technical breakthroughs. Such breakthroughs, or qualitative shifts to new structures of design, do not ordinarily occur at a pop, but become apparent as a line of development reveals its technical potential. From the individual perspective, the breakthrough may be an act of genius that solves a big problem; more often it is the unanticipated result of solving a smaller problem— "overshooting of the mark is characteristic of exploratory activities . . . the size of the discovery need bear no systematic relationship to the size of the initial stimulus" (Rosenberg, 1976: 115). From a systematic perspective, breakthroughs come in terms of structural change within a sector (e.g., a product design shift) or leaps in technology based on the natural organization of matter (mechanical versus chemical versus electronic technologies), whose effects are widely felt. This is where theoretical "science" may reenter the scene, providing, if not the initial breakthrough, the eventual understanding of natural laws on which a long-run technical flowering is based.

Because of the possibility of breakthroughs, technological exhaustion is relative, not final. "It is characteristic of long-term evolution that barriers to growth frequently prove to be temporary" (Sahal, 1981: 69). Dead-ends can be broken out of by structural transformations from one basic design to another, or achieved, pyramid-style, by combining two prior systems and finding the common principle between them. Sahal (1981: 73) gives the example of combining the tractor with the three-point hitch, and calls the principle "creative symbiosis." The same idea can be found in Marx's analysis of combining machines into automatic machine systems or Piaget's analysis of theoretical advances in the sciences (Marx, 1967; Piaget, 1970).

Technical growth is, therefore, a series of waves, not one long one as predicted by the product cycle.[22] In other words, short-term and long-term change are not the same; one is within a structure, the other involves structural transformation.

(6) One can go beyond particular design systems and apply the same principle of structuration to basic technologies or "technological systems" (Freeman et al., 1982; Nelson and Winter, 1977; Rosenberg, 1976). Such systems are not merely linked by inputs and outputs and complementarities, but by common technological and scientific principles, such as those involved in electrical power or control systems. The exploitation of such principles may carry forward a wide range of industries, even lending definition to whole

eras of industrial progress (e.g., mechanical age, electrical age, electronic age). Industries will differ, however, in their ability to exploit such basic technologies (Nelson and Winter, 1977).

TECHNOLOGY AND THE RHYTHM OF ACCUMULATION

The technological determinist views of growth now in vogue are the product cycle for individual industries (Burns, 1934; Kuznets, 1930; Vernon, 1966; Hirsch, 1967) and the wave of innovations view for whole epochs (Mensch, 1979). The evidence does not support either. There is no universal S-curve of growth; sectoral output grows in a variety of patterns (Gold, 1964). Growth paths are characterized by revivals as much as declines, by short-run business cycles as much as long-run patterns. On an aggregate basis, long swings in economic growth bear no obvious relationship to "bunching" of innovations (Freeman et al., 1982).

From what has already been said, what might we expect of the strictly technological contribution to temporal growth patterns? Because technical change is levered, lumpy, and unpredictable, we would expect its effects to be jerky despite the incremental nature of innovative activity or broad structural patterns of development. Nonetheless, structural breakthroughs may open up substantial periods of growth. Such growth might well take a wavelike pattern of upswing and exhaustion of technical change within a distinct design structure in an industry. Because breakthroughs are possible, we would not expect an organic pattern of maturity and decline in most cases, but a process of periodic renewal appearing either as a series of cycles or relatively continuous long-run growth (see Gold, 1964). The chance element in creativity also may lead to clusters of innovations that give shape to a period of growth (Sahal, 1981: 57-60). Finally, technical linkages and common principles across a range of industry means that a major structural breakthrough and period of evolution in one field may trigger a broad front of growth, as is now happening with microelectronics (Freeman et al., 1982; Nelson and Winter, 1977; Rosenberg, 1976; de Bresson and Townsend, 1978). We should not be lulled, however, into forgetting the disjointedness of technical change, the lags and unevenness of application, the odd pathways down which it moves, the multiple waves of innovations, and even the reversals, all of which roil the surface waters and render underlying wave patterns of technical change consistent with a highly incongruous set of events in different industries.

The structural patterns of technology have force in the develop-ment paths of industry simply because production involves physical products, techniques of production, hardware, and a division of labor, and capital must take a "fixed" form as productive and commodity capital (Harvey, 1982). As Rosenberg (1976: 110) says, "the tech-nological level has been more badly neglected than the economist generally recognizes."

Conversely, technical change depends on the other conditions of capitalist growth. From what we know of the process of technical change, we can see the multitude of openings for economic and social influence. The organization of production in profit-guided units linked together by markets obviously means that technical decisions are guided and mediated by market considerations. Moreover, its incremental, practical origins lay technology open to continuous input of cost and price information (David, 1975). Technology requires investment in fixed capital to be installed, so its rhythms bump against those of capital recovery over time. Markets rest on social conditions of consumption, distribution, and division of labor; market saturation, for instance, may be a spur to technical change. Technology is also a means of class struggle between capital and labor.

The rhythms of capital accumulation are themselves prime movers of technical change. Even if innovations are chance events, chances are increased by the degree of effort made, which depends in turn on the rate of investment—and not just R&D. Growth itself means accumulated learning, pressing problems to solve, and the will to tackle them (Sahal, 1981: 110, and references cited there). A notable case is the way postwar Japanese steel makers, pressed by booming growth, broke through supposed technical barriers of scale (Gold, 1979). Indeed, a mere increase in scale through growth demands change in technical structure (including division of labor) of an activity (Sahal, 1981: 65-69; Gould, 1980). Schmookler's (1966) well-known data on innovation actually do not show "market pull" determining technical change, but innovation by learning and adoption by level of investment (Rosenberg, 1976: 260-279). In part, then, growth generates its own technical change; it is self-generative.[23]

In the end, the most fruitful way to look at the problem is in terms of the dialectic between the technical relations of production and value relations, as Harvey (1982: 135) suggests.[24] There is a whole constel-lation of things that must be in place for accumulation to proceed: labor-power, the money and credit system, market institutions, business organization, and so forth. Despite his great attention to technology,

Marx treated it as only one limb of the beast to be studied (Mandel, 1975; Harvey, 1982). Even Schumpeter (1939), considered the father of the technical determinist theory of capitalist growth, actually had an extraordinarily rich conception of the conditions for accumulation, which has been lost in the work of most of his followers.[25] It has been widely noted that periods of growth seem to involve distinct "regimes of accumulation," or growth ensembles (Lipietz, 1977; Gordon, 1978b; Mandel, 1975; Walker, 1981b). To my mind, Schumpeter's (1939) fully articulated analysis amounts to the same thing. The technological system is a fundamental component of such regimes, even if it is not, by itself, the prime mover. It is clear enough how important the railroad and the automobile were to the constellation of conditions of capitalist growth in their respective centuries; or Taylorist and Fordist labor processes; or means of communication such as the telegraph and computerized digital flows. But none of these reduces strictly to a "technology," (i.e., it is institutionalized in other social practices) and none may be said to be *the* source of growth. In short, it is the pattern of accumulation that is the central thread in capitalist growth, not the technology, labor process, or any other single part of the system.[26]

One must also be careful about the image of a structure of accumulation, which implies a rigid functional system of well-fitting pieces, like a tinker-toy. Internal imbalance is a creative force that helps propel the system forward (Harvey, 1982). Thus, in the tension between technology and value lies the potential for both growth and crisis. A better image is one of a system moving along under the force of several driving gears that mesh but one tooth at a time, with much grinding, shaping, spinning, and halting of gears along the way. And, although it can get up speed, it may also grind to a halt as the gears fail to mesh harmoniously under the more and more demanding conditions of rapid growth.

Growth cycles do not turn down, therefore, merely as a result of the diminution of technical change, as epochal technology or big bang theories imply. It might just as well be due to the inability of the system to accommodate an increasing rate of technical change (Harvey, 1982). Downturns also come because of imbalances between sectors, between production and consumption, between labor and capital shares, between money and production. The interesting question is how parts of the system get on the wrong track, out of synch with the others, whether it is a pattern of oil consumption that creates the potential for OPEC's revolt or a pattern of labor bargaining that runs up against the need for automation and wage reductions to meet world competition.

Such contradictions principally develop through the process of over-investment in the wrong avenues of development, whether in too much steel that is no longer needed or in too many nuclear power plants that don't work (Devine, 1980). The imbalances eventually show up in an overaccumulation of one or more forms of capital and lead, ultimately, to crisis if rapid corrections cannot be made (Harvey, 1982).

THE EFFECTS OF TECHNICAL CHANGE

Given the above perspective, we can quickly lay to rest any simple notion of the necessarily beneficial effects of technical change. First, not all technical change is beneficial to capital accumulation. Nuclear power is a prime example. Second, capital accumulation is the issue, not the growth of any other variable, whether employment, income, or personal self-realization, although such growth may accompany accumulation for any number of reasons. It is readily apparent that automation may eliminate jobs, shifts of the division of labor may mean the growth of low-paying jobs, a product innovation in one sector or one country may have its biggest impact elsewhere, cyclical unemployment can destroy people's lives, and so forth.[27] Therefore, the road from innovation to desired social result can be tortuous, if not altogether impassable.

TECHNOLOGY AND SPATIAL PATTERNS OF DEVELOPMENT

We can now add the spatial dimensions of industrial anatomy, technical change, and growth, using the preceding conclusions to guide us and adding new twists that a spatial treatment of accumulation demands. We will see that none of the prevailing views of the regional or locational impact of technical change holds without qualification.

THE SPECIFICITY OF INDUSTRIES AND THE SPATIAL DIVISION OF LABOR

Industries are different and technology plays a big part in those differences, as previously argued. Hence, they have divergent locational needs, whether for a particular material or a particular labor force. As Sayer (1984: 25) puts it, "industry is incredibly complex and differentiated and so too are its products and hence its cost structures, and hence in turn its locational patterns." This complexity is compounded by the divisions among work units within companies and within industries.[28] It makes all universal statements about location patterns immediately suspect, for the primary pattern to grasp is that of a fine mosaic of industrial places.

If industries start from different conditions and follow different development paths because of their technology (as well as their history of labor relations, marketing, etc.), why would we expect them not to start in different places, move differentially, and end up in widely diverse places? All that is required is that the spatial distribution of conditions of production be uneven. This is the valid starting point of Weberian location theory that was lost in the subsequent fixation on transportation and marketing, homogeneous plains, and homogeneous production functions (Losch, 1954; Isard, 1956). But one cannot end with Weberian theory, which takes linkages as given, sees technical change only in terms of factor substitution, treats labor as a one-dimension commodity, and has no theory of economic growth. My position on these topics has already been stated; it will be further amplified as we proceed to consider some theories that go beyond Weberianism in these areas, but still come up wanting.[29]

An issue I have not previously introduced, on which Weberian theory also fails, is the interaction between geography and industry. Weberian models allocate industries to places based on their factor endowments. But there are no such initial conditions once industry is in place.[30] Industry evolves along with places. First of all, industries have a tremendous influence over the spatial distribution of factors of production: they draw labor through migration, they create markets for other industries, they intervene in local politics, and so forth (Piore, 1979; Walker et al., 1981). In other words, industry *produces* industrial space to a considerable degree (Storper, this volume). Geographic unevenness of "factor endowments" is continually recreated by industry's use of space, which affects future location decisions (Massey, 1978, 1983).

Second, preexisting spatial configurations not only steer industry to a spot, they also alter the way industry develops. Geography adds another dimension to industrial evolution. Consider strategy: Companies may seek lower cost either through technical change or by relocating to cheaper labor. The outcomes of such moves become the preconditions of future decisions concerning technology, labor relations, and location. And because technology and labor relations evolve incrementally through experience, such steps are, to a large extent, irreversible; one cannot jump back onto the road not taken (Storper and Walker, 1983). In short, industry growth paths are altered by spatial practice. As a result, one will often find geographically distinct technologies and labor practices within a single industry—especially across national boundaries (Gertler, 1984; Storper and Walker, 1983; Brittain, 1974). At the same time, regional growth paths depend on the way industry develops in places.

Because the technical factor is essential to the character of industries and their locational needs, it should be possible to make some statements about the relation of types of technologies and industry location patterns. One such approach is to use production-based categories such as batch versus continuous flow processing (Storper, 1982; Storper and Walker, 1984). But one must not exceed the limits of what can be said about underlying technological patterns versus empirical geographical regularities. Some limitations are due to the disjunction between structure and outcomes, given intervening contingencies in each particular circumstance; these impede empirical generalization, but not statements about underlying patterns (Sayer, 1982b). Too many things intervene to break the tidy connection between technology and the shape of an industry and between industry structure and its spatial strategy (e.g., the talents of Henry Ford compared to failed car makers who located elsewhere than Detroit). But some of the disjunction involves structural transformation in technology and industry because of their geographic history. The substantial differences between the English and the American automobile industry can be put in no lesser terms.

A case of inappropriate generalization is the search for the golden fleece of the locational patterns of high tech industry. Why should we expect certain shared technological characteristics, such as rapid product development or high automation, to have uniform locational effects across otherwise different industries? Why would electronics in Japan conform to electronics in the United States, given their very different social bases? The search is rendered even more problematic by the uncertainty surrounding the category of high tech itself. Does it refer to product, process, or division of labor? To the rate of technical change? What sort of scale of high and low technology is used? What definition of industry? These may not be insuperable barriers to an analytic meaning for high tech, but the theoretical work is not there as yet.

Are there no broad patterns to the diversity of the spatial mosaic? It will not suffice to end with a plea for specificity alone. Among those who have broken with the neoclassical model of regional specialization of industry and equilization of income, models of spatial hierarchy are popular. Such hierarchies are commonly measured in terms of degree of urbanization, labor skill, sectoral levels (extractive/manufacturing/services), or place in the corporate organization chart (Massey, 1978; Lipietz, 1980; Stanback and Noyelle, 1982; Aydalot, 1981).[31] These models go beyond the Weberian theory of location and its conception of linkages, labor, growth, and technology. In particular, there are three main ways of getting from technology to geographic hierarchy: agglomeration economies, technological maturation, and innovation-diffusion.[32] I will examine each approach at length, and find all wanting, despite obvious areas of application.

AGGLOMERATION

Agglomeration theory is a powerful, if somewhat vague, notion. The idea is that spatial propinquity allows better access to the inputs and outputs necessary for production, thereby raising revenues, lowering costs, and increasing productivity (Hoover and Vernon, 1959; Pred, 1966; Gilmour, 1974). Two aspects of this will be dealt with here: commodity flows (linkages) and labor pooling.[33]

The traditional emphasis in commodity linkage analysis has been on transport costs. Given the friction of distance, proximity gives access to larger markets. This, in turn, allows certain economies of scale in production. While this may have had a certain bearing in the past (Moses and Williamson, 1967), it carries less force today when transport and national markets are so well developed (Storper and Walker, 1984). Yet agglomerations persist. Attention has, therefore, turned to the character of the linkages as well as their mere mass: specialized markets for customized goods, where personal contact and a rapid flow of information are essential (Hoover and Vernon, 1959; Vernon, 1960; Scott, 1982a, 1982b) and technically linked industrial complexes (Castells and Godard, 1974; Norton and Rees, 1979).

Because custom markets and strong technical linkages do not exist for all products, agglomeration cannot be held a universal tendency of industries. Scott (1982a, 1982b) distinguishes two groups of industry: large-scale materials-intensive and small-scale labor-intensive. The former do not depend on agglomeration because they process large volumes of inputs for mass markets, so their linkages are regularized and they get volume discounts on transport. The latter are characterized by a high division of labor, many small exchanges, and customized work. Scott's contribution is to make exchange linkages depend on the character of production, rather than vice versa. It puts technology at the center of agglomeration theory.[34]

But this model will not suffice. For example, sometimes mass production leads to agglomeration, as where there are direct technical linkages (e.g., petrochemical complexes). But the physical ties of chemical pipelines are not the rule. Should we assume that the Detroit auto complex is based on mere physical connections between assemblers and parts suppliers in a linked production system? A territorial complex based on skilled machining is probably a better explanation (Norton and Rees, 1979). On the other hand, mass production (with flexible automation systems) may be consistent with a dispersed production system, as is happening in the Italian auto industry (Murray, 1983). Despite Scott's good efforts, the determining forces of spatial integration-disintegration of the division of labor remain in need of a great deal more theoretical work, after years of neglect.

The geographic scale of agglomeration also has not been carefully addressed. What is the effective range of proximity? If it is "the manufacturing belt" (Norton and Rees, 1979), it borders on solipsism; all distinctions between big cities, systems of big cities, and regional complexes of small towns is lost (see Vance, 1977: 330-35; Pred, 1980; Muller, 1977; Pudup, 1983). Is Silicon Valley a periphery of San Francisco or a core in its own right—or both, depending on what connections we are examining?

Commodity linkages are not enough, of course. Scott tries to tie them in a determinate way to labor demand, arguing that large-scale, materials-intensive industry requires less skilled labor, while small-scale, labor-intensive industry requires more skill. This is the basic formulation of agglomeration economies as regards labor: It takes density to produce an adequate pool of skilled labor, which is in high demand and short supply. Such pools of labor both restrain wage increases and assure supply, and they work to the advantage of workers looking for jobs as well as firms looking for workers. If we take a sufficiently broad view of labor, this pooling of skill among managers, engineers, financiers, and traditional blue-collar skilled workers is a major reason for the agglomerations in Silicon Valley or around Wall Street. But there are cautions here as well.

The treatment of labor demand in agglomeration theory as either skilled or unskilled is reductionist, involving a kind of technical determinism. As previously discussed, labor requires a richer conception of performance, control, and reward (Storper and Walker, 1983). Problems of labor control and cost, especially under a regime of collective bargaining, may make wage demands higher in urban areas because of militancy and "pattern bargaining," leading some industries to seek out dispersed locations. Storper (1984) suggests that the state of labor relations in Brazil generates a more concentrated pattern than in the United States. On the other hand, labor control may be best secured in dense settlements, such as immigrant ghettoes. The definition of skill is also more problematic than it first appears. The so-called unskilled at the periphery may not be a usable workforce if they haven't sufficient industrial discipline.

Nor can labor and linkages easily be joined in a technically determinate way. Part of the reason for the agglomeration of electronics firms in Silicon Valley is the vigorous web of linkages among component suppliers and product assemblers; this seems to depend strongly on the type of final products, such as computers. In cases in which there are few linkages, such as the space communication systems of being built in Melbourne, Florida, there is little agglomeration, despite the presence of a huge pool of technical labor (Glasmeier, this volume).

Agglomeration theory has the virtue of simultaneity; that is, it goes beyond the partial equilibrium of Weberian theory.[35] Agglomeration theory touches on the active creation of spatial organization by industry. But it does so in a very limited way, through contiguity, not strategy. It misses the possibility of action at a distance and of the active intervention of business in labor markets, local politics, and the like. Industry does not depend only on the passive action of the market, distance, and the cumulative weight of many industries to make places suitable for its purposes.

At the same time, agglomeration theory has a limited conception of feedback from spatial organization to industry. It usually includes some notion of diseconomies of agglomeration, which is a catch-all for such things as rising labor costs, militancy, and traffic congestion. Although there is no question that such things happen, and may lead to industrial decentralization (Saxenian, 1983), this weak formulation gives little indication why and when such effects take hold. Moreover, both agglomeration and disagglomeration are normally seen as simply shifting conditions of factor supply, rather than as altering the shape of technical change and industrial development itself.

TECHNOLOGICAL MATURATION AND RELOCATION

It is frequently argued that technical change results in the dispersal of workplaces from core industrial areas.[36] I will consider two reasons for such decentralization.[37] The first theory of decentralization is that the maturation of production processes—greater standardization and mechanization—breaks the bond of dependence on skilled labor and allows industry to seek out cheaper, unskilled labor supplies in peripheral regions (Hoover, 1948; NRPB, 1943; Vernon, 1960, 1966; Hirsch, 1967; Norton and Rees, 1979). While many times this does happen, there are so many exceptions that it does not serve as a valid generalization. Both the major process and the intervening causes need to be formulated more carefully.

First, the idea rests on an organic notion of growth and maturation of technology that needs to be sharply circumscribed. Technical change does not unfold in such a simple, universal, and unproblematic way, for all the reasons previously discussed. Some industries never move beyond the craft or small batch stage and remain tied to skilled labor, for example, medical instruments (Hekman, 1980b). Some go looking for cheap labor precisely because they must achieve mass production without mechanization, for example, U.S. semiconductor assembly in the Far East. At the same time, the Japanese took the route of mechanization for better quality control and did not disperse production. Sometimes industries revert to more customized production because of product innovations, for example, the recent bloom of custom microprocessor producers in Silicon Valley.

Second, labor demand does not follow so easily from technique. Increased automation may lead to a type of reskilling and recentralization, as is likely for U.S. semiconductor assembly. Specialized skills developed by a long-standing industry-labor relation, rather than skill in general, may be the issue (Sahal, 1981: 59). Why else do certain industries, such as jet engines, cling to particular centers despite considerable production change over time (Storper, 1982)? Moreover, the demand for labor does not reduce to skill. Relocation may occur for reasons of labor control without technical change, although ordinarily both control and technical change are involved in decentralization (NRPB, 1943; Massey and Meegan, 1982). Conversely, successful labor control can mean the avoidance of technical change and/or relocation.

Third, the supply of labor is not given in the fashion assumed by the product cycle model. The distribution of skills is not a simple gradient from a skilled center to an unskilled periphery. In the past, the northeast and midwest United States were dotted with industrial towns making everything from shoes to farm implements and benefitting from the influx of skilled European immigrants (Lindstrom, 1978; Pudup, 1983). On the other hand, some of the cheapest labor has always been found in the central cities. Then, of course, labor can migrate to or with production. Skilled labor often is relocated along with new plants (Pratt, 1911). Differential mobility of labor segments makes the decision more complex: Medium-range skilled technicians are often the least mobile and most sought after labor force (Oakey, 1979).

Finally, both the active recruitment policies (Piore, 1979) and the ongoing employment practices of industry (Storper and Walker, 1983) shape the local labor force. Conversely, the experience industry has with labor and other local conditions shapes its development path; production does not dictate location in a strictly linear way.

Textiles are the classic case of relocation to cheap unskilled labor in the South. Hekman (1980a) makes a good case for a product cycle interpretation of this shift. Nonetheless, there are difficulties. Mechanization advanced for 50 to 100 years before it resulted in the kind of deskilling the model anticipates. Even then, some skilled workers were exported to the South along with the new plants: Why could this not have been done earlier? And Hekman treats the designation unskilled as completely unproblematic, as if there were nothing to making agrarian folk into industrial workers (Thompson, 1967). Carlson (1981) opposes this view, arguing that Southern textile makers tended to cluster in areas where industrial experience was greatest. In fact, the industry had built a base in the South going back to the 1840s and grew rapidly by 1860—long before the technical maturation Hekman argues for. Did slavery have any affect on labor force and industrial development in the South before 1865 (Genovese, 1967)? Why did many skilled workers who set up operations in the South fail before 1880? Nor is labor control in the North, rather than simply skill, given more than passing consideration (Hekman, 1980a: 711). Why were a series of labor forces, from mill girls to Greeks, brought in over the years if labor cost and control were not already an issue in New England? Finally, it will not do to treat technical change as essentially complete by 1910, as Hekman does. Why hasn't the synthetic/knit revolution resulted in a locational shift as profound as the one circa 1900? It appears that one must take a closer look at why capitalists gave up on the workers of New England after a century and found a suitable replacement in the South at the time they did. They may have even pushed technical change to accommodate this strategy. In any event, technical change may have made the shift possible, but it cannot bear the full weight of explanation.

The second explanation of decentralization is the lessening pull of interindustry linkages on large, standarized operations. The parallel between delinking and deskilling is commonly observed (NRPB, 1943; Hoover, 1948; Vernon, 1966; Hirsch, 1967; Hekman, 1980a). Scott (1982a, 1982b) again tries to tie the two together in terms of production changes. Mass production means greater integration of formerly scattered units and greater throughput of materials; hence, larger, more regularized markets and bulk discounts. Industries thus move from small-scale, labor-intensive to large-scale, materials-intensive.

But, for all the reasons previously cited, we must be cautious about imputing such a pattern to all industries. Products, especially capital goods, do not all mature to mass consumption items. Process does not evolve single-mindedly to mass production. So all industries cannot be expected to go through delinkage—even if they begin in an agglomerated state. More troublesome to the delinking hypothesis is the occasional evidence of *increased* agglomeration accompanying greater integration and scale of production, as in the case of automobiles in Detroit in the 1920s (NRPB, 1943: 101) and textiles near Boston in the 1810s through 1840s (Vance, 1977: 333). Is it a result of the elimination of more widely dispersed smaller competitors (the opposite of Scott's scenario of large plants driving out more agglomerated small firms)? Or is it that large plants in some industries have powerful agglomerative effects, drawing around themselves small suppliers and such (Oakey, 1983; Glasmeier, this volume)? That depends on the linkage characteristics of the particular industry, which varies by technology and social practice in production, product, and division of labor. One cannot simply read linkages off from production technique. Note, for example, the reconcentration of some American industries (e.g., Caterpillar tractor), production methods unchanged, in order to imitate the Japanese practice of Kanban, or rapid delivery of parts without holding large inventories.[38]

Hekman (1980a) adds that the crucial link in New England textiles was to the producers of textile machinery, who needed to be in close contact with customers and who relied on skilled labor more than textile makers themselves. Only as machine making became standardized and mechanized was it possible to break this link and move textiles south (often keeping machine making in the North). In light of the evidence about machine makers and technical change, this is a powerful argument (and used to good effect against the water power site theory of textile location). Nonetheless, one wonders why the changes in machine making, which were widespread, did not lead to similar moves south among a wider spectrum of industries. Again, it must be said that we have little theoretical guidance as to how to grasp these variations in the development of markets, linkages, and agglomeration/disagglomeration.

INNOVATION AND GROWTH

A major school of thought sees growth as the simple product of technical innovations, which spring from urban-industrial agglomerations and diffuse from there (for reviews, see Pred, 1973; Malecki, 1983). This raises four issues: the site of innovation, where innovations are taken up, the effects of innovation, and the role of innovation in growth as a whole.

Innovation springs chiefly from industrial practice, not spatial proximity in general. While industrial concentrations may generate a great deal of innovation, rapid technical change also occurs in industries with a more dispersed location, such as farm implements (Pudup, 1983). The rate and character of innovation depends heavily on the industrial base, and there is no necessary relation between innovative capacity and agglomerative location pattern. The link is usually made on the basis of concentrations of skilled labor or information, but the practical limits of products and production have more impact on innovation than skill or knowledge in general. As Chinitz (1961) observes, Pittsburgh's heavy industry complex colors, and limits, the possibilities of innovation. Similarly, the Midwest, as a center of metal working, is most likely to generate innovations of this type, rather than be a center of innovation in general, as Norton and Rees (1979) imply. Tests of the "incubator" hypothesis indicate that innovation (new firms, new growth) takes place in central cities chiefly when that is already the preferred locus of the industry (Struyk and James, 1975; Leone and Struyk, 1976; Cameron, 1980; Nicholson et al., 1981).

To the extent that innovations flow from organized R&D work—which is uncertain—they will appear in the preferred locations for such work. While R&D is highly centralized, it also has specialized labor demands by industry and linkages depending on whether it is market, process, or basic science oriented (Malecki, 1981: 315, 317; Gold, 1979).

Clearly, however, innovation does not just reinforce existing practice and is not necessarily used in the same place it originates. There is a chance element in the spatial as well as the temporal incidence of innovation and its spread. But systematic influences also operate. The markets and linkages of an industry have a great effect on the paths down which change travels. They particularly affect the potential for spinoff products and firms; the multiple component and market niche character of computers makes them a prime spinoff generator (Glasmeier, this volume).

More generally, the industrial environment of an established center, in terms of the domination of labor, capital, and linkages by one industry, may stifle the development of new activities in any number of ways, as Chinitz says of Pittsburgh.[39] The search for an environment of its own, a clean slate—especially as regards labor—is probably a basic reason behind the tendency of so many new industries, from automobiles in Detroit to semiconductors in Santa Clara County, to seek out new pastures for development. Once there, they reproduce the conditions of their own growth for a time (Storper and Walker, 1983; Saxenian, 1983; Christopherson, 1982; Scott, 1984). This is a much more far-reaching reason for industrial dispersal than is technical maturation. Its result is a temporal sequence of places developed in concert with particular industries or industrial complexes (Watkins, 1977; Gordon, 1978a; Walker, 1981). The mosaic of unevenness thus rests in time as well as space.

Of course, the identification of technical change with local growth is too easy. As the literature on branch plants has amply documented, the beneficial effects of growth may be felt far away (e.g., Pred, 1977; Erickson and Leinbach, 1979). This is not because branch plants are lacking in technical change (e.g., semiconductor plants in Phoenix) (Sayer, 1984), but because the effects of technical change depend on the labor force, type of process, linkage patterns, and so on. For example, process change may reduce employment at a site, or it may spur relocation; product innovation may prompt investment in a wholly new line, produced at a different plant.

Finally, a shifting division of labor within industries may be the most important technical change as far as spatial patterns of growth are concerned. Not many places can link their star to the microelectronics industry, with its rich harvest of spinoffs; indeed, net new firm formation is a tiny percentage of overall growth (Malecki, 1981). For most places, the prospect is of capturing a limb or two of the industrial system. In the complex mosaic of industrial location, almost every facility is a branch plant. That applies even to R&D labs, which are normally not the places where the growth spurred by their innovations occurs. Yet, regardless of what R&D labs actually produce, they are good things for local growth because industry is investing a lot in them and hiring high-paid labor to work there. In other words, it is probably more important to attract investment in expanding segments of industry than to spawn new industries. The benefits of capitalist growth depend on where surplus value ends up, not where technical change begins.

At a larger scale, where regions develop on the basis of a diverse industrial system, growth has broader and deeper roots than technical innovation, contrary to the view of Norton and Rees (1979). Self-generative regional development involves the generation, retention, and investment of surplus value, the development of a capable, diverse workforce, the interaction of a range of industrial activities, the creation of a suitable social and physical infrastructure, an energetic and growth-oriented state apparatus, and the like. That is, it requires a workable growth ensemble. A few places achieve something like this ideal—although no region is completely self-contained and not every place that is broadly suitable will, in fact, sustain growth. All of the famous examples of rapid regional growth in the United States—early nineteenth century New England, the late nineteenth century Midwest, twentieth century California, late twentieth century Houston—had long years of prior development, including agriculture, mercantile trade, and accumulation, a home market, government promotion, and various humble industries such as food processing (see, for example, Pred, 1966; Pudup, 1983; Platt, 1983). Although they benefitted in many ways from the antecedent growth of more advanced regions, they did not depend primarily on branch plants or technological spinoffs for their growth. Capitalism is not monolithic and it has shown its capacity to generate growth, including technical innovation, on fresh soil on numerous occasions.

CONCLUSION

In sum, technology has a significant effect on industrial location and the course of regional development, particularly in the way it shapes the distinctive character and growth path of industries. Nonetheless, we must eschew technical determinism as a mode of analysis. Neither logic nor evidence sustain most prevailing theories of broad spatial patterning based on technological forces. Too many other variables intervene in the open system of the space-economy for sweeping generalizations about the effects of technical patterns to be borne out. We need a more supple approach to technology and its geographic effects. Geographers must, therefore, employ a structural-realist mode of analysis that takes cognizance of technology as one of the basic structuring forces of the industrial system, but puts it in proper relation to other elemental structures of capitalism (such as class and competition), allows for human agency and the many contingencies of history, and inserts space into the process of industrial development (see Sayer, 1983). In short, geography cannot be read off from technology.

NOTES

1. I do not take up the important questions of market and business organization or of the circulation of money and capital, except peripherally.

2. Ordinary neoclassical economics has no technological determinism because technology is not regarded as an interesting problem. Choice of techniques is a static process of weighing relative prices of labor and capital. This Hicksian view of price—induced factor substitution, in which technique (and capital equipment) is treated as perfectly malleable, has been exposed to withering criticism (Kennedy and Thirlwall, 1972). From one side, it has been argued that capitalists do not cost minimize by choosing one factor over another (Salter, 1966); besides, technical change is lumpy and balancing factor prices is a pipedream (Rosenberg, 1976: 61-84). From another side, it has been argued that the price of capital is a fiction, which cannot be known without prior knowledge of technical coefficients and the distribution ratio between capital and labor (Hunt and Schwartz, 1972; Harcourt, 1972). For my part, I have always wondered why, in a neoclassical world with perfect substitution, homogeneous production functions, and perfect markets, all industries do not converge on an identical technique. And, indeed, in his attempt to defend neoclassical capital theory, Samuelson ended up having to assume a world in which all industries have the same labor-capital ratio (Harcourt, 1972).

3. Technology typically refers only to products and processes of production. This begs the question of organization of production and circulation, except where it sneaks in the back door by reference to the "technical division of labor" or "market"—as if either were not the problematic result of human institution and choice. In this chapter I continue an extended argument for the importance of the division of labor as a structural category of economic analysis that cannot be reduced either to process/product, class, market, or corporations (Walker, 1985a, 1985b). The division of labor must be included, along with both market and corporate forms of organization, among the whole range of the forces of production. The organization of production and circulation is very much structured by technical problems of transforming nature and rendering it useful (and accessible) to people, independent of the structuring influence of class and power (the social relations of production). I will often use the word "technology" to encompass the division of labor and organization of production as a matter of convenience, given the clumsiness of the phrase "forces of production."

4. The distinction between use and use-value parallels that between exchange-value (price) and value (see Harvey, 1982).

5. A slightly different case is the resource-based industries, which have certain problems of extracting resources from the earth in common and, therefore, share some developmental characteristics (Markusen, 1983; Perloff, et al., 1960).

6. It should be clear that what has been said in the previous section about technical linkages, imbalances, and innovation applies both within and between industries.

7. Von Tunzelman's (1978) study of steam power in the industrial revolution vindicates Marx's analysis.

8. Although the notion of deskilling has had, since Adam Smith, a certain purchase among mainstream economists (see NRPB, 1943).

9. Marx also touches on two dimensions of the labor process that need greater development: the transfer of materials from one step of processing to another (Storper, 1982) and the monitoring of process, movement, and product quality (Bright, 1958). These

become increasingly important with continuous flow, integrated machine systems, and assembly-line processes and with electrical/electronic control systems, or what commonly comes under the heading of "automation." In general, we need a better grasp of chemical and electronic process improvement, comparable to the treatment of mechanical systems.

10. Of course, there are also generalizable principles involved in the physical processes of production—which is why one can speak of "technological systems" based on epochal technologies and families of machines. But these are not as widespread as the principles of division of labor or mechanization.

11. It is by no means obvious at any time which of several available techniques is best; indeed, even the physical efficiency of any process is hard to grasp in a single, comparable measure (Gold, 1979).

12. There is also a good deal of unevenness in the mechanization and automation of various steps in production (Bright, 1958; Kinnucan, 1983).

13. The labor-saving bias is augmented by considerations other than cost, of course, especially the preference of capitalists for docile machinery over refactory people.

14. Piore (1968) found a consistent class bias of engineers against labor-intensive methods. But he also emphasizes the priority of technological imperatives over careful weighing of labor considerations in the development of production.

15. The following rests heavily on my reading of Rosenberg (1976: especially 189-212, 108-125; 1982: 55-80), David (1975), Nelson and Winter (1977), Sahal (1981), Gold (1976, 1979), and Piore (1968).

16. I use the term "science" here in the way it is commonly understood in the literature, as theoretical laboratory research. If the term is broadened to mean the systematic use of mechanical, chemical, or electronic principles, rather than relying on the traditional knowledge of the worker about a craft, then, of course, science is applied constantly in modern industry (Marx, 1967; Rosenberg, 1976: 126-138; 1982: 141-163).

17. Indeed, excessive social stratification and lack of communication between high-level engineers and lower-level people often blocks technical change.

18. There are, however, considerable problems in dating innovations (Rosenberg, 1976: 72; Freeman et al., 1982: 45-51; Gold, 1979).

19. There is a common misconception that processes are mechanized of a piece, when they are, in fact, unevenly developed; highly sophisticated technology may rub elbows with wholly manual operations (Bright, 1958; Kinnucan, 1983; Piore, 1968).

20. Such effects are felt in the price sphere as well, even where there are no direct technical complementarities at work.

21. Within a basic structure of product and production there is likely to be an evolution of the labor process toward greater mechanization if markets are expanding (Abernathy and Utterback, 1978).

22. And they are often overlapping because old products in the same industry do not necessarily die out as the new come in (e.g., blenders versus cuisinarts).

23. However, full capacity utilization does not always appear to be conducive to experimentation, so innovations may come earlier and later in the business cycle (Mansfield, cited in Sahal, 1964: 118).

24. Only the net can be cast more widely to include use-value relations of commodity consumption and work, in so far as these cannot be strictly reduced to either technology or value.

25. Mandel has also sometimes been interpreted as a technological determinist, wrongly I think.

26. Although we should be able to rank the overall importance of different parts.

27. On the other hand, one should not leap to hasty conclusions about cause and effect, given the many technical and nontechnical sources of interaction in the system; automation of manufacturing, for example, need not lead to permanent unemployment if it stimulates additional product demand through falling cost, generates linked demands for materials and equipment, and is accompanied by investment of surplus profits in other sectors.

28. Having already made the point about the difficulty of defining industries, I need only say here that the word "industry" is used in full cognizance of the problems.

29. My target here is not directly Weberian theory, but most of what follows applies to it as well. For more on Weberian/neoclassical themes, see Walker (1981), Walker and Storper (1981), and Storper and Walker (1983).

30. Nor before, given the social workings of preindustrial capitalism or other modes of production.

31. There are serious problems of specification in most such models that I cannot pursue here. The measures are often poorly theorized (e.g., "services"—see Walker, 1985b) and the frequently assumed correspondence among different measures by no means always obtains (e.g., between urban and corporate hierarchy—see Pred, 1977).

32. In my view, product cycle theory is a way of tying the three together, not a separate theory. I am currently working on a full critique of the product cycle.

33. Innovation, which is included in some models of agglomeration (e.g., Pred, 1966, 1973), is considered separately below. Other agglomerative forces, such as finance and information, are beyond the scope of this chapter.

34. A similar connection is implied, if not explicit, in the work of Hoover (NRPB, 1943; Hoover and Vernon, 1959), who is also very sensitive to the importance of changing marketing arrangements (e.g., via merchant houses) as well as volumes. However, some would argue that the organization of commerce is more important to agglomerations than product characteristics or technical linkages.

35. It thus avoids such foolishness as attributing all location to market pull, when industry is its own biggest market. It also speaks to a world in which raw materials are a small portion of inputs and of diminishing influence on location (Perloff et al., 1960).

36. It is usually assumed that such relocation starts from an initial agglomeration, although that is by no means always true.

37. Other factors, such as availability of space in cities, cannot be dealt with here (see Walker, 1981). I agree with Scott (1982a, 1982b) that they are secondary to the matters discussed here.

The main neoclassical theory of decentralization is that transportation costs have fallen and flexibility has increased via the truck. Although improved means of circulation have undoubtedly made greater dispersal possible, they have not determined it (Walker et al., 1981; Storper and Walker, 1983; Scott, 1982a, 1982b). It all depends on what kinds of transport needs industry has. The demand for transportation has an effect in the timing, if not the shape, of the technology of transport (Walker, 1978). Even Chinitz, who gave the truck theory of dispersal its major statement, soon cast doubt on it (see Chinitz, 1960, 1961).

38. My thanks to Phil Shapira for pointing this out.

39. Not only do changing industrial conditions demand changing regional circumstances, but the costs of past growth start to catch up with capital: labor militance, governmental regulations, deteriorating infrastructure, rising housing costs, and so on (Saxenian, 1983; Walker, 1981a).

REFERENCES

ABERNATHY, W. and J. UTTERBACK (1978) "Patterns of industrial innovation." Technology Review 80.

ARONOWITZ, S. (1978) "Marx, Braverman, and the logic of capital." Insurgent Sociologist 8: 126-146.

AYDALOT, P. (1981) "The regional policy and spatial strategy of large organizations," pp. 173-185 in A. Kuklinsky (ed.) Polarized Development in Regional Policy. The Hague: Mouton.

BHASKAR, R. (1978) A Realist Theory of Science. Atlantic Highlands, NJ: Humanities Press.

BRIGHT, J. (1958) Automation and Management. Cambridge, MA: Harvard University School of Business Administration.

BRITTAIN, J. (1974) "The international diffusion of electrical power technology, 1870-1920." Journal of Economic History 34: 108-121.

BURNS, A. (1934) Production Trends in the United States. New York: National Bureau of Economic Research.

CAMERON, G. (1980) "The inner city: new plant incubator?" pp. 351-366 in A. Evans and D. Eversley (eds.) The Inner City. London: Heinemann.

CARLSON, L. (1981) "Labor supply, the acquisition of skills and the location of southern textile mills, 1880-1900." Journal of Economic History 41(1): 65-73.

CASTELLS, M. and J.L. GODARD (1974) Monopolville. Paris: Maspero.

CAVES, R. (1980) "The structure of industry," pp. 501-545 in M. Feldstein (ed.) The American Economy in Transition. Chicago; University of Chicago Press.

CHINITZ, B. (1961) "Contrasts in agglomeration: New York and Pittsburgh." American Economic Review, Papers and Proceedings: 279-289.

——(1960) Freight in the Metropolis. Harvard, MA: Harvard University Press.

CHRISTOPHERSON, S. (1982) "Family and class in a new industrial city." Ph.D. dissertation, University of California, Berkeley.

CRESSY, P. and J. MACINNES (1980) "Voting for Ford: industrial democracy and the control of labor." Capital and Class 11: 5-33.

DAVID, P. (1975) Technical Choice, Innovation, and Economic Growth. Cambridge: Cambridge University Press.

de BRESSON, C. and J. TOWNSEND (1978) "Notes on the inter-industrial flow of technology in postwar Britain." Research Policy 7: 48-60.

DEVINE, J. (1980) "Overinvestment and cyclic economic crises." Ph.D. dissertation, University of California, Berkeley.

ELGER, T. (1979) "Valoration and deskilling: a critique of Braverman." Capital and Class 7: 58-99.

ERICKSON, R. and T. LEINBACH (1979) "Characteristics of branch plants attracted to nonmetropolitan areas," pp. 57-78 in H. Seyler (ed.) Nonmetropolitan Industrialization. New York: Winston/Wiley.

EWEN, S. (1976) Captains of Consciousness. New York: McGraw-Hill.

FREEMAN, C., J. CLARK, and L. SOETE (1982) Unemployment and Technical Innovation. Westport, CT: Greenwood Press.

GENOVESE, E. (1967) The Political Economy of Slavery. New York: Vintage Books.

GERTLER, M. (1984) "Regional capital theory." Progress in Human Geography 8(1): 50-81.

GIDDENS, A. (1979) Central Problems in Social Theory. Berkeley: University of California Press.

GILMOUR, J. (1974) "External economies of scale, interindustrial linkages, and decision making in manufacturing," pp. 335-362 in F.E.I. Hamilton (ed.) Spatial Perspectives on Industrial Organization and Decision Making. New York: John Wiley.

GINTIS, H. (1972) "Consumer behavior and the concept of sovereignty." American Economic Review 62(2): 267-278.

GOLD, B. (1979) Productivity, Technology, and Capital. Lexington, MA: Lexington.

———(1976) "Tracing gaps between expectations and results of technological innovations: the case of iron and steel." Journal of Industrial Economics 25(1): 1-28.

———(1964) "Industry growth patterns: theory and empirical results." Journal of Industrial Economics 13: 53-73.

GORDON, D. (1978a) "Up and down the long roller coaster," pp. 22-35 in U.S. Capitalism in Crisis. New York: Union for Radical Political Economics.

———(1978b) "Class struggle and the stages of urban development," pp. 55-82 in A. Watkins (ed.) The Rise of the Sunbelt Cities. Beverly Hills, CA: Sage.

GOULD, S.J. (1980) The Panda's Thumb. New York: Norton.

HARCOURT, G. (1972) Some Cambridge Controversies in the Theory of Capital. Cambridge: Cambridge University Press.

HARVEY, D. (1982) The Limits to Capital. Oxford: Basil Blackwell.

HEKMAN, J. (1980a) "The product cycle and New England textiles." Quarterly Journal of Economics 94(4): 697-717.

———(1980b) "Can New England hold onto its high-technology industry?" New England Economic Review (March/April): 5-17.

HIRSCH, S. (1967) Location of Industry and International Competitiveness. Oxford: Clarendon.

HOOVER, E. (1948) The Location of Economic Activity. New York: McGraw-Hill.

———and R. VERNON (1959) Anatomy of a Metropolis. Cambridge, MA: Harvard University Press.

HUNT, E.K. and J. SCHWARTZ (1972) A Critique of Economic Theory. Baltimore: Penguin.

ISARD, W. (1956) Location and Space Economy. New York: John Wiley.

JEWKES, J., D. SAWERS, and R. STILLERMAN (1959) The Sources of Invention. New York: St. Martin's.

KENDRICK, J. (1973) Productivity Trends in the United States, 1948-69. New York: National Bureau of Economic Research.

KENNEDY, C. and A. THIRLWALL (1972) "Surveys in applied economics: technical progress." Economic Journal 82: 11-72.

KINNUCAN, P. (1983) "Flexible systems invade the factory." High Technology 3(7): 32-42.

KUHN, T. (1970) The Structure of Scientific Revolutions. Chicago: University of Chicago Press.

KUZNETS, S. (1930) Secular Movements in Production and Prices. Boston: Houghton-Mifflin.

LEBOWITZ, M. (1977/1978) "Capital and the production of needs." Science and Society 41(4): 430-448.

LEONE, R. and R. STRUYK (1976) "The incubator hypothesis: evidence from five SMSAs." Urban Studies 13: 325-331.

LINDSTROM, D. (1978) Economic Development of the Philadelphia Region, 1810-50. New York: Columbia University Press.

LIPIETZ, A. (1980) "Interregional polarisation and the tertiarisation of society." Papers on the Regional Science Association 44: 3-17.

——(1977) Le Capital et Son Espace. Paris: Maspero.

LOSCH, A. (1954) The Economics of Locations. New Haven, CT: Yale University Press.

MAK, J. and J. WALTON (1978) "The great productivity surge in western river transport." Journal of Economic History 8: 12-45.

MALECKI, E. (1983) "Technology and regional development: a survey." International Regional Science Review 8(2): 89-126.

——(1981) "Science, technology, and regional economic development: review and prospects." Research Policy 10(1): 312-334.

MANDEL, E. (1975) Late Capitalism. London: New Left Books.

MANSFIELD, E. (1972) "Contribution of R&D to economic growth in the United States." Science 175(February): 477-486.

MANWARING, T. and S. WOOD (1984) "The ghost in the machine: tacit skills in the labor process." Socialist Review 74(14/2): 57-86.

MARKUSEN, A. (1983) Profit Cycles, Oligopoly, and Regional Transformation: Working Paper 397. Berkeley: University of California, Institute of Urban and Regional Development.

MARX, K. (1967) Capital. New York: International Publishers.

MASSEY, D. (1983) "Industrial restructuring as class restructuring." Regional Studies 17(2): 73-89.

——(1978) "Regionalism: some current issues." Capital and Class 6: 106-125.

——and R. MEEGAN (1982) The Anatomy of Job Loss. London: Methuen.

MENSCH, G. (1979) Stalemate in Technology. Cambridge, MA: Ballinger.

MOSES, L. and H. WILLIAMSON (1967) "The location of economic activity in cities." American Economic Review, Papers and Proceedings: 211-222.

MULLER, E. (1977) "Regional urbanization and the selective growth of towns in North American regions." Journal of Historical Geography 3(1): 21-39.

MURRAY, F. (1983) "The decentralization of production—the decline of the mass-collective worker." Capital and Class 19: 74-99.

National Resources Planning Board [NRPB] (1943) Industrial Location and National Resources. Washington, DC: U.S. Government Printing Office.

NELSON, R. and S. WINTER (1977) "In search of a useful theory of innovation." Research Policy 6: 36-76.

NICHOLSON, B. M., I. BRINKLEY, and A. EVANS (1981) "The role of the inner city in the development of manufacturing industry." Urban Studies 18: 57-71.

NORTON, R. D. and J. REES (1979) "The product cycle and the spatial decentralization of American manufacturing." Regional Studies 13: 141-151.

OAKEY, J. P. (1983) Industrial Growth and New Firms. San Francisco: Harper.

OAKEY, R. P. (1979) "Labour and the location of mobile industry: observations from the instruments industry." Environment and Planning A 11: 1231-1240.

PERLOFF, H., E. DUNN, E. LAMPARD, and R. MUTH (1960) Regions, Resources, and Economic Growth. Baltimore: Johns Hopkins University Press.

PETERS, T. (1983) "The mythology of innovation, or a skunkworks tale." Stanford Magazine (Summer/Fall): 13-21.

PHILLIPS, A. (1971) Technology and Market Structure: A Study of the Aircraft Industry. Lexington, MA: D. C. Heath.

PIAGET, J. (1970) Structuralism. New York: Basic Books.

PIORE, M. (1979) Birds of Passage. Cambridge, Cambridge University Press.

————(1968) "The impact of the labor market upon the design and selection of productive techniques within the manufacturing plant." Quarterly Journal of Economics 92: 602-620.

PLATT, H. (1983) City Building in the New South. Philadelphia: Temple University Press.

PRATT, E. E. (1911) The Industrial Causes of Congestion of Population in New York City. New York: Columbia University Studies in History, Economics, and Public Law.

PRED, A. (1980) Urban Growth and City Systems in the United States, 1840-60. Cambridge, MA: Harvard University Press.

————(1977) City Systems in Advanced Economies. London: Hutchinson.

————(1973) Urban Growth and the Circulations of Information, 1790-1840. Cambridge, MA: Harvard University Press.

————(1966) The Spatial Dynamics of Urban Growth in the United States, 1800-1914. Cambridge, MA: MIT Press.

PRICE, W. and L. BASS (1969) "Scientific research and the innovative process." Science 164: 802-806.

PUDUP, M.B. (1983) Packers and Reapers, Merchants and Manufacturers: Industrial Structuring and Location in an Era of Emergent Capitalism. Berkeley: University of California, Department of Geography.

ROSEGGER, G. (1979) "Diffusion and technological specificity: the case of continuous casting." Journal of Industrial Economics 28(1): 39-53.

ROSENBERG, N. (1982) Inside the Black Box: Technology and Economics: Cambridge: Cambridge University Press.

————(1976) Perspectives on Technology. Cambridge: Cambridge University Press.

SAHAL, D. (1981) Patterns of Technological Innovation. Reading, MA: Addison-Wesley.

SALTER, W. (1966) Productivity and Technical Change. Cambridge: Cambridge University Press.

SAXENIAN, A. (1983) "The urban contradictions of Silicon Valley: regional growth and the restructuring of the semiconductor industry." International Journal of Urban and Regional Research 7(2): 237-262.

SAYER, A. (1984) "Industry and space: a sympathetic critique of radical research." Society and Space.

————(1983) "Theoretical problems in the analysis of technological change and regional development," pp. 59-73 in G.J.R. Linge (ed.) Spatial Analysis, Industry, and the Industrial Environment. London: John Wiley.

————(1982a) "Explanation in economic geography: abstraction versus generalization." Progress in Human Geography 6(1): 68-88.

————(1982b) "Explaining manufacturing shift: a reply to Keeble." Environment and Planning A 14: 119-125.

————(1981) "Defensible values in geography: can values be science free?" pp. 29-56 in R.J. Johnston (ed.) Geography and the Urban Environment, vol. 4. London: John Wiley.

SCHERER, F.M. (1970) Industrial Market Structure and Economic Performance. Chicago: Rand McNally.

SCHMOOKLER, J. (1966) Invention and Economic Growth. Cambridge, MA: Harvard University Press.

SCHUMPETER, J. (1939) Business Cycles. New York: McGraw-Hill.

SCOTT, A. (1984) "Territorial reproduction and transformation in a local labor market: the animated film work of Los Angeles." Society and Space 2: 277-307.

————(1982a) "Production system dynamics and metropolitan development." Annals of the Association of American Geographers 72(2): 185-200.

————(1982b) "Locational patterns and dynamics of industrial activity in the modern metropolis." Urban Studies 19: 111-142.

SHEPHERD, W. (1970) Market Power and Economic Welfare. New York: Random House.

STANBACK, T. and T. NOYELLE (1982) Cities in Transformation. Totowa, NJ: Rowman and Allenheld.

STORPER, M. (1984) "Who benefits from industrial decentralization? Social power in the labour market, income distribution, and spatial policy in Brazil." Regional Studies 18(2): 143-164.

————(1982) "The spatial division of labor: technology, the labor process, and the location of industries." Ph. D. dissertation, University of California, Berkeley.

————and R. WALKER (1983) "The theory of labor and the theory of location." International Journal of Urban and Regional Research 7(1): 1-44.

STRUYK, R. and F. JAMES (1975) Intrametropolitan Location of Industry. Lexington, MA: D. C. Heath.

THOMPSON, E. P. (1967) "Time, work discipline, and industrial capitalism." Past and Present 38: 56-97.

USELDING, P. (1974) "Studies of technology in economic industry," pp. 159-219 in P. Uselding (ed.) Research in Economic History, Supplement 1. Greenwich, CT: JAI Press.

VANCE, J. (1977) This Scene of Man. New York: Harper & Row.

VERNON, R. (1966) "International investment and international trade in the product cycle." Quarterly Journal of Economics 80(2): 190-207.

————(1960) Metropolis 1985: An Interpretation of the Findings of the New York Metropolitan Region Study. Cambridge, MA: Harvard University Press.

VON TUNZELMAN, G. N. (1978) Steam Power and British Industrialization to 1860. Oxford: Clarendon Press.

WALKER, R. (1985a) "Class, division of labor, and employment in space." in D. Gregory and J. Urry (eds.) Social Structure and Spatial Relations. Cambridge: Cambridge University Press.

————(1985b) "Is there a service economy?" Science and Society.

————(1981a) "A theory of suburbanization: capitalism and the construction of urban space in the United States," in M. Dear and A. Scott (eds.) Urbanization and Urban Planning Under Advanced Capitalist Societies. New York: Methuen.

————(1981b) "Industrial location policy: false premises, false conclusions." Built Environment 6(2): 105-113.

————(1978) "The transformation of urban structure in the 19th century United States and the beginnings of suburbanization," pp. 165-213 in K. Cox (ed.) Urbanization and Conflict in Market Societies. Chicago: Maaroufa Press.

————and M. STORPER (1981) "Capital and industrial location." Progress in Human Geography 5(4): 473-509.

————and E. WIDESS (1981) "Performance regulation and industrial location: a case study." Environment and Planning A 13: 321-338.

WATKINS, A. (1977) "Uneven development within the U.S. system of cities." Ph. D. dissertation, New School for Social Research, New York.

12

Technology and Spatial Production Relations: Disequilibrium, Interindustry Relationships, and Industrial Development

MICHAEL STORPER

☐ WHY DO MAJOR UPHEAVALS in industrial technologies and the location patterns and regional development paths that accompany them seem to take geographers, planners, and economists by surprise? Whether it be the decline of industrialized, high-income regions in advanced capitalist nations, the rise of new capitalist challengers such as Japan, the "miracles" in Korea and Brazil, the subsequent setbacks in Brazil, or the rise of new industries in altogether new types of spatial milieux, technological and spatial change is much less regular and often more dramatic than most of our theories would predict.

This chapter proposes a heuristic for analyzing the historical evolution of spatial production relations of industries. The purpose of this exercise is to establish the theoretical grounds on which we might begin to account for dramatic increases in the spatial capabilities of industries, based on the notion that we also must account for dramatic changes in technologies themselves. To do this, we must engage debates in both economics (technology) and location theory (spatial consequences of innovations); this chapter is a modest attempt to use some of the conceptual tools that have been developed in these fields to consider dramatic and long-run historical changes in technologies and location.

To get at this interaction between the process of locational change in particular industries, via particular innovations and strategies, and the overall development of the forces of production and the locational capabilities contained therein, I want to retain a focus on the factors of production—supply and demand—that are necessary for any activity to locate in a particular place. The focus on the spatial demand for and supply of factors of production normally directs our attention to price-cost considerations and the private and singular nature of decisions in capitalist economies. In neoclassical models in which equilibrium is the real or tendential reality, analysis focuses only on private and singular constrained optimization problems, and aggregate supply and demand take care of themselves and produce the context within which further private and singular decisions may be taken. As we shall see, this is not a sufficient basis for economic history; once we reject the microeconomic foundations of the neoclassical problematic, we reject the identity it proposes of microeconomics and macroeconomics. This raises the need for another level of analysis entirely: We must supply some additional mesolevel analytics that are missing from neoclassical paradigms and not well developed in political economy approaches (Storper, 1984b). In this view, the macroeconomy is built from, and microeconomic decisions are taken in light of, the existence of time- and space-differentiated intermediating systems of social practices, such as industries, markets, and nations. This chapter concentrates on firms and industries as mediating and differentiated systems. Analyzing these systems is key to the relationships between private and singular decisions—the domain of neoclassical theories—and general and structural outcomes—the traditional domain of Marxism. The relationships can be viewed as the unfolding of a "structuration" process in which there is more structure to singular decisions than in neoclassical theories, but in which there is room for human agency; and in which humanly-constructed systems of practice, such as industries, are still appropriated and influenced by overarching structural forces. These dialectics converge at the mesoanalytical level.

Instead of rejecting the traditional microeconomic categories of supply and demand, I resituate them at the center of the structuration approach to economic history. Factors, in this usage, denote all the relationships required for production (including, for example, markets) and I seek an understanding of the transformations in these relationships by the mesolevel systems that comprise the economy. The facility (as a concrete entity) and the economy (as a series of dynamic historical relationships) in which it is situated produce each other. We therefore must ask: How do demands for and supplies of locationally important factors of production change historically?

What regulates the changes in these relationships in production and the locational surface? In seeking explanation of these changes, I attempt a more developmental approach to industry location, a more locational view of regional development, and a more industrial perspective on the development of the forces of production.

LOCATION AND PRODUCTION: SUPPLY- VERSUS DEMAND-ORIENTED PARADIGMS

Existing theories and research traditions concerning industry location and regional development, on one hand, and those interested in long-term growth, technological innovation, and development, on the other, are neither well-integrated with one another nor individually well-equipped to deal with long-run, nonparametric changes in growth and location. This gap is especially apparent regarding growth of a "qualitative" nature; that is, involving epochal changes in the productivity of labor and the use of territory.

For example, the strongest tradition of industry location research rests on neoclassical production theory. The location decision is merely one of a series of decisions about the mix and substitution of factors for one another along a given production function, arriving at the optimal factor mix and location simultaneously (Isard, 1956; Moses, 1958). Common to all work within the neoclassical perspective is the notion of the firm as a cost-minimizing or revenue-maximizing "decision taker": It does not have the power to influence or create the conditions for its own survival under an equilibrium structure. Even if it is conceded that the conditions for general equilibrium do not exist, as in the case of joint products and scale economies, the theoretical principles can be applied so as to view locational decisions as guided by marginal substitutions and rational expectations. Implicitly, the locational analytics of these approaches emphasize the configuration of factor supplies in space, and they can be labeled "supply-based" location theories. There are no substantive analytics for showing how the demand for factors from production could influence the evolution of supplies in any direction. Any changes in direction of demand that could create a directional bias in factor supply markets (this is not to be confused with the outcome of adjustments through spatial interaction/trade) are exogenous to a given equilibrium tendency: These include changes in tastes and preferences; changes in technology; and changes in population, income distribution, or income levels.

Much regional science has accepted neoclassical production theory as a basis for industry development and location; its self-appointed task has then been to concentrate on the region as an input-output system and to model the intraregional and interregional interactions that lead to equilibrium patterns of commodity trade and factor prices (Isard, 1969). Regional development consists of industry-locational surface interactions guided by neoclassical production dynamics. Newer regional science models, such as the product cycle, allow for contrary factor intensities by allowing a greater role for some developmental dynamics, such as Schumpeterian innovation, labor quality differentials, economies of scale and joint products, amortization periods on fixed capital stock, and learning curves in production and marketing. Still, with its assumption of competitive price taking across the product cycle, the vision is retained of regional development as an orderly hierarchical filtering process (see Markusen, 1983, 1984).

A second major perspective on industry location has arisen in the last few years, principally as a response to the neoclassical theories. It focuses on the changing organization of production and investment priorities. Industries undergo "restructuring" periodically. In this view, the investment decision dominates the location decision (different from the optimizing world in which the two are identical at the margin; see Walker and Storper, 1981).

The restructuring school has tended to take as its point of departure the existence of the macroeconomic crises of capitalism, whether in the form of the short business cycle (Kuznets cycle) or the long wave of capital accumulation (Kondratieff wave) (Kondratieff, 1979). Industry restructuring occurs in light of these broader forces, which are a reflection of the necessary internal relationships of capitalism; that is, the law of value and the movements in the social average rate of profit. The point of departure is thus quite the opposite of the neoclassical theories' essential concern with the private and singular.

In industrial restructuring research, the concern with the macroeconomic and necessary is reflected in the focus on the process of cutback and rationalization. In Massey and Meegan's extensive research (Massey and Meegan, 1978, 1982; Massey 1978a, 1978b), for example, industrywide and worldwide dynamics are antecedent to factor-supply conditions in generating the impulses toward locational shifts in investments; firms did not move simply in response to factor-supply differentials that had, after all, existed for some time. These decisions are strongly geared to macroeconomic conditions:

The corporate survivors of each crisis and merger wave in the economy are a new breed in terms of scale and organization. Their investments are bearers of new products and processes, whereas obsolete processes and uncompetitive plants and firms disappear. Industrial restructuring means alterations in product mix, production technique, and organizational structure. These events control factor demands and trigger spatial reorganization of production to secure them, and open up new opportunities to exploit preexisting factor supply differentials (Massey and Meegan 1978, 1982; Bluestone and Harrison, 1982; Walker and Storper, 1981). The basic approach of restructuring analysis to date is industrial organization, rooted in revenue-maximization assumptions. Class and social conflict, the role of the state, corporate and market power, and nonoptimal behavior are central elements in economic decision making (Massey and Meegan, 1982; Frobel et al., 1980; Bluestone and Harrison, 1982).

Arguably, the process of cutback is more directly derivable from the necessary internal relationships of the capitalist macroeconomy than would be technological innovation and growth, and so the concentration on it is understandable. It is the producers who can no longer participate in the formation of industry or social average profit rates who are physically eliminated from the capital stock of the industry and the economy (but see Bluestone and Harrison, 1982, for counterexamples within this general trend). Recently, Markusen (1983, 1984) and Sayer (1984) have begun to advance the restructuring argument beyond cutbacks to a consideration of innovation and growth. As in the research on rationalization, however, Markusen concentrates on the notion that there is a structural rationality and temporal regularity to the nonneoclassical dynamics of growth, just as there are to those of decline. Uneven development, in this sense, is "normal" in a capitalist industrial system.

I do not intend to take up the complex debate over whether there is some predictability to dramatic technological and spatial transformations. Rather, I want to consider the process by which new products, new production methods, and new territories are incorporated into the space economy. Given that there is a general imperative to accumulate capital in capitalism, what are the sources of the specific technological and spatial innovations that make accumulation possible? The argument here is that when we assert the existence of a macroeconomy that develops in crisis-bound disequilibrium, we imply that it contains many "openings" for dramatic changes in its parts, from firms to industries to the collective forces of production, through the exercise of human agency at all these levels. Firms, industries, and territories are complex systems of social practice and sites of social action: The "forces of production," which are an abstraction of these particularities, are an outcome.

In the next section, I demonstrate what a "bottom up," but still structural, view of the evolution of technologies and spatial production relations into macroeconomic forces of production and territorial development might look like. It is only a starting point for research, and much more detailed theoretical elaboration is needed. First, we will reconsider some microeconomic concepts, allowing us to see the relationships between the short-term strategies of firms with respect to products, processes, and locations. These generate long-term paths for industries and regions by simultaneously excluding other paths in real time, possibly blocking other paths that are more factor-optimal over the long run. Second, we can see how at any moment the macroeconomy is composed of technically, socially, and locationally diverse concrete industries. These are the intermediating systems to which I alluded above. Third, the social forces of production are developed from complex interindustry interactions. Macroeconomic structure (movements in the social average rate of profit, and physical input-output characteristics) is the "totalization" of these smaller structuration processes (the term is defined by Sartre, 1963).

THE DEVELOPMENT OF INDUSTRIES AND THEIR FACTOR DEMANDS

We can begin at the microeconomic level with firms and their choices of production techniques. Contrary to neoclassical theory, the potential profits from shifts in production techniques and outputs are not guided by continuous marginal substitutions under conditions of factor- and product-price competition (Shaikh, 1981). Much recent work in Marxian, and neo-Ricardian economics provides a solid intellectual foundation for historical conceptions of industry evolution. It does this by showing that technologies are heterogeneous physical capital, for which money value (and, therefore, price-quantity) cannot be determined a priori, but only after overall prices are determined and net revenues are divided into profits and wages. This division is determined "before" the value of capital is set, via a distributional struggle between capital and labor. This means that industries do not have the foresight or the physical flexibility to establish equilibrium production techniques and to keep them in equilibrium from one moment to the next. Instead, the evolution of industries involves a historical process of investment and qualitative change; although this process may achieve some alteration in factor proportions in response to trends in prices, this by no means determines the course of an industry's development because the trends in those prices themselves are determined through active human agency (i.e., by the prevailing income distribution) (Gertler, 1984). The importance of these theories is that they create a formal opening for considering the evolution of industry organization and technologies as qualitative, uneven, historical processes rather than movements from one equilibrium state to another (see Kaldor, 1972; Robinson, 1953, 1956; Garegnani, 1970; Pasinetti, 1977; Harvey, 1982; Scott, 1980).

Firms take actions, such as product innovation or change of production technique, that may not correspond to neoclassical rules regarding factor intensities or prices (this applies both to the continuous substitution and rational expectations varieties). In the neoclassical view, short-term events of this type must be considered mistakes that should be eliminated in intertemporal competition. In the disequilibrium view, short-term superprofits can be translated into long-term advantage.

Superprofits are the difference between the firm's profit rate and the sector's average. At a broader level, they may be the difference between the sectoral average and the social average (Mandel, 1975). They arise because, in an economy in which adjustments are lagged considerably in space and time due to the heterogeneous nature of capital and its valuation only ex post, the average state of production technique and profit rate (dubbed by Shaikh, 1981, the sector's or economy's "regulating capital"), may be quite different from that of the most innovative or successful producers. Until the conditions of production are further socialized, the firm will enjoy this differential between its profit rate, the sectoral average, and the intersectoral social average (Mandel, 1975). This logic can apply to normal profits as well. Clark and Gertler (1984) point out that firms undertaking the "spatial innovation" of locating in nonunion areas will be able to lay off workers during cyclical downturns. As capital consists of heterogeneous capital goods, most technical coefficients and prices of production are fixed in the short run. Total production costs can only be adjusted by varying the quantity of output. This requires laying off workers, as capital cannot be laid off at will. Only those firms with the ability to adjust the labor input will make normal profits; others will go out of business in spite of identical production functions. After an industry shakeout, the next growth period will tend to be characterized by the behavior of the survivors and this will set the course of development of the sector. The long-term course of the sector is built up from transitory events (Kalecki, 1972).

This logic applies more directly to spatial innovations associated with technological innovations than is commonly realized, in two ways. First, the technological innovations that generate superprofits may directly make possible new forms of spatial behavior by altering the traditional determinants of industry location: scale of production, division of labor/cooperation, interplant or interindustrial linkages, and so on. Second, if a firm or industry is making superprofits, it may have the flexibility to undertake radically new spatial strategies that are not constrained by strict price-competition. That is, it will have more tactical flexibility with respect to production costs and, therefore, location. Out of this there may be unexpected invasions of new territories or spatial restructuring in sectors, much more dramatic than what one would expect with a conventional equilibrium price approach. In contrast to the neoclassical image of incremental spatial expansion, I employ the metaphor of leapfrogging.

Innovation of new products (new use-values, as opposed to new production techniques exclusively) brings out the above logic even more clearly (see Sayer, 1984, for a related discussion). To see this, we have to think about why new products arise and what regulates their acceptance in society. Obviously, these are the stuff of which many volumes are made, and it is impossible to do them justice here. In the neoclassical view, new products might arise from exogenous shifts in tastes and preferences and their associated utility functions. Alternatively, they must compete favorably with preexisting products for the consumer's income, which is totally and efficiently allocated in equilibrium. New products represent the household's version of substitution, either along given utility functions or by new functions for old ones, with concomitant devaluation of the old ones. But this doesn't fit the concrete circumstances of technologically dynamic economies. In high-income economies (but usually with a not-too-unequal income distribution) there are resources available to find new occupations, and income is distributed widely enough to make innovation a paying proposition. Given the existence of enough nonsubsistence income and enough people with it, and the potential for unequal profit and growth rates in industries, there are myriad opportunities for inventions to gain acceptance. These opportunities are not regulated exclusively by rational substitutions among fully allocated household incomes.

Moreover, the temporally and spatially specific patterns of material life in which new use-values arise are not merely material representations of prices, whether equilibrium or not. For example, many historians of technology seem to agree that new products, as well as new processes, emerge in response to some perceived problem or opportunity in a concrete situation (see Landes, 1969; David, 1975; Rosenberg, 1979; Piore, 1968). The emergence of new technologies, then, is related to how they fit existing and expected patterns of material life, and these patterns vary considerably from place to place according to the development of particular political economies. To say that utility functions change exogenously is wrong from this perspective; they are endogenous because they emerge from temporally and spatially specific habits of daily life (see Bourdieu, 1978; Giddens, 1982; Pred, 1982).

The existence of extraordinarily high profit rates, along with very high initial prices for new products, has been demonstrated conclusively by Markusen (1985). These short-term profits are related to the long term of the industry or product sector concerned in the same way as we described above for process innovations; although profit and growth rates may decline through the process of interfirm and interindustry profit rate equalization, later in the industry's development path the successful product configurations will have had important effects on the industry's long-term evolution.

We can point to two major types of long-term effects of early battles over the configurations of use-values. First, the firms successful in the short term may have a strong effect on the product configuration for a given use-value and this, in turn, will have strong effects on the process technology employed during the rest of the industry's development. The struggles currently being waged in electronics are prime examples of this process, for those that win now will gain market shares and barriers to entry will arise that block new products due to subsequent price-competition. There will be important locational effects of the outcome of such a struggle within a given use-value.

Second, this framework may help us to understand why new industries tend to locate in new places or in new types of social-spatial production complexes. In the neoclassical framework, external economies in central regions overwhelm the advantages of lower prices for some factors in peripheries (the so-called Hicks-neutral advantages). New industries should not only form territorial production complexes (agglomerations), but they should do so in the cities, and the release of resources to new cities should be orderly and gradual as producers shift production functions and consumers move along their utility functions. Empirically, however, developed regions actually do not seem to perform their innovative seedbed roles. Instead, new industries seem to form their initial production complexes in relatively undeveloped (though not underdeveloped) regions without deep (or at least without similar) industrial histories (Watkins, 1982, shows this in a study of urban growth rates and industry mixes for U.S. metropolitan areas). Industries producing new use-values or radically new processes have their own locational requirements that are unlike those of previously existing industries (Rosenberg, 1970; Storper, 1982). How are these historically unique factor requirements created in areas that are not at the center of the urban-regional hierarchy? Why are they not necessarily transmitted down the hierarchy (see Pred, 1977)? The discussion above suggests that these new industries must have some way of attracting the factors of production they need, thereby causing territorial production complexes to come into being in regions that did not previously possess those factors. It seems that new industries wield a kind of factor attraction power due to their high profit rates and consequent abilities to pay certain factors (the relatively expensive ones), in addition to taking advantage of the frequent slackness of allocation of relatively cheap factors in capitalist economies (cheap implies abundant in absolute terms in a macroeconomic disequilibrium, as in Marx's case of the reserve army of labor). New industries may indeed agglomerate, but not in the places suggested by conventional models. Competitive conditions eventually may assert themselves, causing new industries to decentralize, but this does not erase the development that has occurred in the meantime, nor the damage to the regions that were excluded, as new industries avoid the habits of labor, capital, consumers, and the respective institutions (unions, entrepreneurship, class alliances, consumption-income structure, and ways of daily life) of already specialized regions. Additionally, new industries often develop new spatial forms within regions, such as the high tech integrated research and manufacturing complex, the greenfield site before it, and the company town further back in time.

Thus far we have observed that firms (and people) are agents of innovations in the technological and spatial patterns of development of industries. Ultimately, the process of interfirm competition socializes the successful products, technologies, and spatial behaviors among the firms that remain in an industry. What were once innovations become the sectoral average—are incorporated into its regulating capital—and, in turn, are replaced by other innovations. The temporal paths of process innovations have been observed by product cycle theorists (Vernon, 1966, 1979; Norton and Rees, 1979; Hoover, 1948; Krumme and Hayter, 1975; Thomas, 1975). Markusen (1983, 1984) has made a significant advance over conventional versions of the industry development cycle that were rooted in competitive cost taking. She observes that whole industries have differentiated profit rates at the various stages of their development. At any moment, the economy is composed of industries for which profit rates are higher or lower than the social average, even if they are intensely internally competitive.

High profit stages are characterized by oligopoly. Oligopoly stems from discontinuous technical opportunities; that is, by the lack of continuous entry. This amounts to a condition of blocked intersectoral capital mobility (Sylos-Labini, 1969; Storper, 1984a). This condition of disequilibrium means that entire industries, at certain points in their histories, can exercise their economic power with respect to the rest of the economy by inventing new forms and destroying old forms of spatial organization. This is similar to the possibility, suggested above, that new industries and firms can produce new places for location or new spatial forms; but here it applies, as well, to the standardization-decentralization phase of the industry's development. The point is that existence of a certain regularity in an industry's development (from centralized to decentralized, external to internal economies, and so on) does not imply a similar regularity with respect to the specific patterns of centralization-decentralization they undertake. The spatial succession is not that regular because these industries have power with respect to the rest of the economy.

With this sense of the jaggedness of the development of particular industries, their factor demands, and their spatial capabilities, we can move up another level to the macroeconomy and, specifically, to the forces of production. The latter concept is particularly important, and so "its various meanings must be carefully defined and distinguished from common misconceptions" (Harvey, 1982: 99). Forces of production are not the same as technology. Technology refers to the tools and machines used in production, the physical design of production processes, the (technical) detail division of labor, the actual employment of labor powers, and the particular methods of coordination, control, and hierarchy used in production (Harvey, 1982). Productive forces refers to all of these, but embeds them within a general notion of the power to transform and appropriate nature through human labor, not reducible to specific devices because, at some point, quantitative change becomes qualitative change.

Let me endeavor to show why it is that the form and content of these qualitative changes are built from the actions taken by human agents in firms and industries (which allow them some freedom of action due to the condition of disequilibrium). This view can first be contrasted with notions that the course of development of the forces of production reflects a structural response to aggregate conditions. In the neoclassical view aggregate movements in relative factor prices are reflected in the direction of technological change in the economy as a whole (see Habbakuk, 1962, on the British-American capital intensity/mechanization debate). In some equilibrium interpretations of Marx's economics, as well, aggregate tendencies in the ratio of surplus to necessary labor time are reflected in an overall tendency to replace labor with fixed capital (Roemer, 1980; Okishio, 1961). These analyses imply, incorrectly, that the mere fact of aggregate factor prices or movements in value/productivity ratios and the social average rate of profit functionally determine the concrete evolution of the forces of production. In the view advocated thus far, the forces of production—both in their physical aspects (products and processes) and in their economic representations (value magnitudes and price levels)—are built from the actions of specific firms and industries. (Note that Marx, 1967: 383-384, understood this tendency to be built from specific industries.)

Marx and Ricardo both paid considerable attention to the tendency toward technological dynamism in capitalist economies as a whole and the effects of interindustry competition in tending to equalize profit rates over time. We can see now that this tendency only exists because its opposite reigns most of the time: Industries' profit rates are usually differentiated. The macroeconomic effect of such profit rate variations is intersectoral capital flows with simultaneous overcapitalization of some industries (those with superprofits) and disinvestment from others. Innovations in some sectors that permit them superprofits have broader effects in the economy: They create an incentive to renew other industries that do not have as high a rate of profit and are losing capital as a result.

Fundamental changes in the technologies of particular industries transform the forces of production generally. New branches of production arise to serve developing industries, as in the case of the shipping industry serving textile production in the nineteenth century, steel production and ship building arising to serve it in turn. Office-based industries have emerged in the twentieth century to serve growing corporate and financial functions and they, in turn, have generated a tremendous demand for electronic information technologies. The powers of production created by the development of these new branches of production are not exclusively available to the leading edge industries, however; they can effect across-the-board changes in the productive power of the economy, as is now occurring with the transition from mechanical to electronic means of guiding production.

The specific technological and organizational advances of particular industries, then, enter into a totalizing process in which they are themselves subsequently transformed into something supraindividual. They begin as materially concrete phenomena, but they have effects in the value and price spheres and, in turn, this provides incentive for them to be diffused physically. The process of competition then begins again. But its concrete form comes from below.

At certain junctures in industrial history it appears that these changes combine into qualitative shifts in the productivity of labor, the organization of capital, and so on. These major renewals are said to be the source of long waves (Kondratieff waves) of capital accumulation (see Mensch, 1975; Freeman et al., 1982; Kondratieff, 1979) or of capitalism's creative destruction (Schumpeter, 1934). It is not important, for our purposes, whether the temporal regularity attributed to this process by adherents of the long wave is correct or not. Most students of technology agree that major renewals do occur (see Rosenberg and Frischtak, 1983). At these points epochal developments of the forces of production as a whole can be said to have occurred. The transition from artisanal to factory production (from manufacture to machinofacture) and its concomitant advance from steam to electric power is one such epoch. The current transition from mechanically structured, dedicated production technologies to electronically guided, flexible systems is another. The convergence of these physical events logically has convergent effects in the price and value spheres. When newly productive industries that are interdependent are relatively scarce, they grow and the social average rate of profit will tend to move upward; the regulating capital for the whole economy is affected. Conversely, when ensembles of industries reach points at which their conditions of production are widely socialized, reflected in excess capacity and stagnant productivity norms, they enter a process of rationalization together and the social average rate of profit will tend to fall. The point here is that this macroeconomic fact does not move on its own volition; the content and magnitudes of its movements, and their spatial manifestations, are built from the actions of many firms and industries in disequilibrium. Most of the time that capital accumulation is occurring, capitalist economies are characterized by firms and industries with differentiated profit rates. Most of the growth in capitalist economies is achieved via this disequilibrium, because the tendency toward equilibrium (profit rate equalization) is approached only during short-term rapid-growth periods and rapid-decline periods.

Other qualitative dimensions of macroeconomic change can now be envisioned. New industries and new branches of production in the social division of labor have their own productivity norms, spatial capabilities, and consequent wage-income structures. As the locations of specific branches of production are rearranged, they alter the industrial mixes of territories and, consequently, they alter the occupations of the populations of those territories and the levels and distributions of income in those regions and nations. The outcome of these dynamics and the social struggle over income distribution set up a definite pattern of consumption that, in turn, has much to do with the evolution of the industries themselves (i.e., whether they become mass producers or luxury producers, and so on). There is nothing inevitable about the structures of markets.

Thus, as social labor is redivided into different branches due to the interindustrial sharing and relinking around new forces of production, those industries effect a new price system. Here we come full circle in our critique of the neoclassical perspective: Instead of the price system determining the new equilibrium allocation of resources among sectors, the initial evolution of those sectors, the strategies of firms therein, and the totalization of those sectoral changes in the form of qualitatively renewed forces of production in general produces the new price system from the bottom up. This new system of prices and production-consumption relations is not produced automatically by any structural imperative. It represents a historically achieved and temporary regime of accumulation, which is a system of macroeconomic relations that allow accumulation to proceed for a period of time (Lipietz, 1983; Aglietta, 1976).

This perspective should not be misread as an endorsement of the neoclassical view, in which the macroeconomy is not an important theoretical construct because it is assumed that localized price-taking behavior sums to the macroeconomy. In effect, the macroeconomy has no independent, supraindividual structure. In the restructuring view advanced here, the macroeconomy is composed of all manner of time- and space-differentiated systems, including industries, markets, and nations. Through its system of prices, inputs-outputs, production/consumption/income relations, and social average rate of profit, the macroeconomy does indeed exercise some coordination among industries and produces pressures on the decisions and strategies undertaken therein. But the interrelationships among the parts are only roughly coordinated over the evolutionary course of the whole (in Marxist parlance, the mesolevel systems are "relatively autonomous"). This emphasis on the interactions between time- and space-differentiated systems of social practice in the structuration of the whole, in my view, is consistent with Marx's attempt to advance a richer theory of capitalist competition than that which has been proposed by the neoclassicals, in which there are only choices made by individuals, albeit under complex constraints to optimization. In my conception, there is a complex strategic war, involving conflict and cooperation, individuals and collectivities; economies are actively constructed, not passively evolved. Manifestations of the development of the forces of production—technological innovation, organizational growth, new spatial strategies—are leading-edge structural changes that take precedence over the adaptive changes that are at the heart of the neoclassical model. Changes in the form and organization of production are evidence of changes in the rules of the game. What does this emphasis on demand mean for location and space economy? If the economy is constructed through the exercise of human agency, and not inevitable, much the same can be said for space. Put another way, economic space can be seen, logically, as produced through the exercise of human agency within structural conditions of disequilibrium.

DEMAND AND SUPPLY INTERACTIONS

The production of space can serve only as a starting point for analyzing the spatial development process: Taken in its pure logical form, it implies that there is no locational problematic, as industries simply produce the space they need. This is where the difference between a theory or logic of historical causality and a model of the unfolding of historical process becomes critical. Ultimately, it is necessary to develop an analysis of the relationship between changes in factor demands and the status of factor supplies, and then to see the ways that use of those factor supplies by industries influence their further evolution, and so on, in a continuous historical stream.

A useful heuristic for integrating the demand-oriented, historical theory of the development of the forces of production, with the demand and supply-oriented process in a common spatial framework, is to divide the constellation of criteria for locating economic activities into locational capabilities and locational specifications. Locational capabilities are the same as the conditions enabling geographic generalization of industry. They can take explicitly spatial forms, as in the development of more effective modes of transportation and communications, or they can take ostensibly nonspatial forms, as in the internalization of supply linkages, scale economies, or creation of oligopolistic markets. Locational capabilities reflect the evolution of the forces of production over time. The industries that comprise these forces of production, that have these capabilities, demand factor supplies.

Of course, to the extent that such resources that are demanded are not costless or ubiquitous, all capabilities define locational specifications, particularly under conditions of price competition (if not immediately, then eventually, in the development of any sector). Thus, even industries with increasing capabilities have a compulsion to choose specific locations carefully and sometimes even to change them without a major change in capabilities. Locational specifications describe the particular attractions of sites that are the spatial distribution of conditions requisite to making profits or superproifts in an industry.

A whole field of analysis is opened up, then, for analyzing the capability-specification relationship, which has been taken to be unproblematic in neoclassical and allied approaches. Unfortunately, I do not have the space here that would be required to elaborate these spatial interactions more fully, and so I will conclude by simply summing the theoretical opening I have tried to create here.

CONCLUSION

Factor demand and factor supply in the production of space must be treated as internal components of one model and based on a nonequilibrium, demand-led concept of the development of production. Demand is not determined by supply, because the overall development of the forces and relations of industrial production continually alter the basis of the space economy, regardless of the initial distribution of factor supplies. Industrial development guided by the heterogeneity of capital goods, the perversity of factor demands, and the social struggle over wages and profits—involves changes in the absolute availability and relative importance of factor demand and supply relationships that specify industry locations. This means not only a different landscape to describe, but changing logics of industry location in the course of development. Strategies for making profits undergo qualitative changes within and between industries and territorial political economies over time, as the basic competitive constraints and social possibilities change. There is a strong degree of irreversibility in the ensuing adjustments, and previous economic and locational configurations are irrevocably lost.

In addition to pointing out this historical dimension to the evolution of spatial production relations, I have attempted in this chapter to make some suggestions about the theoretical tools we need to explain such changes. In particular, I have endorsed a notion of economic structure that is at once more cognizant of the suprain-dividual nature of economic forces, and yet provides a greater role for human agency in the construction of particular economic outcomes, than approaches that take as their point of departure either market processes or macroeconomic structural aggregates. In so doing, I have suggested that the nature of innovations and the nature of spatial changes associated with them is likely to be much less predictable and much more dramatic than we are led to believe in conventional approaches, whether neoclassical spatial succession or structurally determined spatial rationalizations in the course of Marxist long waves. This should focus our attention on the role that politics can play in determining these outcomes on which our lives depend.

REFERENCES

AGLIETTA, M. (1976) The Theory of Capitalist Regulation: The U.S. Experience. London: New Left Books.

BLUESTONE, B. and B. HARRISON (1982) The Deindustrialization of America. New York: Basic Books.

BOURDIEU, P. (1978) Outline of a Theory of Practice. Cambridge: Cambridge University Press.

CLARK, G. and M. GERTLER (1984) A Kaleckian Model of Regional Production. Chicago: University of Chicago, Department of Geography.

DAVID, P. (1975) Technical Choice, Innovation, and Economic Growth. Cambridge: Cambridge University Press.

FREEMAN, C., J. CLARK, and L. SOETE (1982) Unemployment and Technical Innovation. London: Frances Pinter.

FROBEL, F., J. HEINRICHS, and O. KREYE (1980) The New International Division of Labor. Cambridge: Cambridge University Press.

GAREGNANI, P. (1970) "Heterogeneous capital, the production function, and the theory of distribution." Review of Economic Studies (July): 407-436.

GERTLER, M. S. (1984) "Regional capital theory." Progress in Human Geography 8(1): 50-81.

GIDDENS, A. (1982) A Contemporary Critique of Historical Materialism. Berkeley: University of California Press.

HABBAKUK, H. J. (1962) American and British Technology in the Nineteenth Century: The Search for Labor-Saving Inventions. Cambridge: Cambridge University Press.

HARVEY, D. (1982) The Limits to Capital. Chicago: University of Chicago Press.

HOOVER, E. M. (1948) The Location of Economic Activity. New York: McGraw-Hill.

ISARD, W. (1969) General Theory: Social, Political, Economic, and Regional. Cambridge, MA: MIT Press.

———(1956) Location and Space Economy. New York: John Wiley.

KALDOR, N. (1972) "The irrevelance of equilibrium economics." Economic Journal 82(December): 1237-1255.

KALECKI, M. (1972) Selected Essays on the Dynamics of the Capitalist Economy, 1933-1970. Cambridge: Cambridge University Press.

KONDRATIEFF, N. D. (1979) "The long waves in economic life." Review (Spring).

KRUMME, G. and R. HAYTER (1975) "Implications of corporate strategies and product cycle adjustments for regional employment change," in Collins and Walker (eds.) Locational Dynamics of Manufacturing Activity. New York: John Wiley.

LANDES, D. (1969) The Unbound Prometheus. Cambridge: Cambridge University Press.

LIPIETZ, A. (1983) Les Transformations dans la Division Internationale du Travail: Considerations Methodologiques et Esquisse de Theorisation. Paris: CEPREMAP, 8302.

MANDEL, E. (1975) Late Capitalism. London: New Left Books.

MARKUSEN, A. R. (1985) Profit Cycles, Oligopoly, and Regional Development. Cambridge, MA: MIT Press.

———(1983) Profit Cycles, Oligopoly, and Regional Transformation: Working Paper 397. Berkeley: University of California, Institute of Urban and Regional Development.

MARX, K. (1967) Capital, vol. 1. New York: International Publishers.

MASSEY, D. (1978a) "Regionalism: some current issues." Capital and Class 6: 106-125.

———(1978b) "Capital and locational change: the U.K. electrical engineering and electronics industry." Review of Radical Political Economics 10(3): 39-54.

———and R. MEEGAN (1982) The Anatomy of Job Loss. London: Methuen.

———(1978) "Industrial restructuring versus the cities." Urban Studies 15: 273-288.

MENSCH, G. (1976) Stalemete in Technology. Cambridge, MA: Ballinger.

MOSES, L. (1958) "Location and the theory of production." Quarterly Journal of Economics 73: 260-269.

NORTON, R. D. and J. REES (1979) "The product cycle and the spatial decentralization of American manufacturing." Regional Studies 13: 141-151.

OKISHIO, N. (1961) "Technical changes and the rate of profit." Kobe University Economic Review 7: 85-99.

PASINETTI, L. (1977) Lectures on the Theory of Production. New York: Columbia University Press.

PIORE, M. (1968) "The impact of the labor market on the design and selection of production techniques in the manufacturing plant." Quarterly Journal of Economics 92: 602-620.

PRED, A. (1982) Place as Historically Contingent Process. Berkeley: University of California, Department of Geography.

———(1977) City Systems in Advanced Economics. London: Hutchinson.

ROBINSON, J. (1956) The Accumulation of Capital. London: Macmillan.

———(1953) "The production function and the theory of capital." Review of Economic Studies 21: 81-106.

ROEMER, J. (1980) "A general equilibrium approach to Marxian economics." Econometrica 48(2): 505-530.

ROSENBERG, N. (1979) "Technological interdependence in the American economy." Technology and Culture (January): 25-50.

——(1970) "Economic development and the transfer of technology: some historical perspectives." Technology and Culture (October): 550-575.

——and C. R. FRISCHTAK (1983) "Long waves and economic growth: a critical appraisal." American Economic Review 73(2): 146-151.

SARTRE, J. P. (1963) The Search for a Method (H. Barnes, trans.). New York: Knopf.

SAYER, A. (1985) "Industry and space: a sympathetic critique of radical research." Society and Space 3.

SCHUMPETER, J. (1934) Business Cycles. New York: Harper & Row.

SCOTT, A. J. (1980) The Urban Land Nexus and the State. London: Pion.

SHAIKH, A. (1981) "The poverty of algebra," pp. 266-300 in I. Steedman (ed.) The Value Controversy. London: Verso.

——(1980) "Marxian competition versus perfect competition: further comments on the so-called choice of technique." Cambridge Journal of Economics 4(1): 75-83.

——(1978) "Political economy and capitalism: notes on Dobb's theory of crisis." Cambridge Journal of Economics 2: 233-251.

STORPER, M. (1984a) "Against spatial structure: history in economic geography, part I." Los Angeles: University of California, Graduate School of Architecture and Urban Planning. (unpublished)

——(1984b) "Against spatial structure: history in economic geography, part II." Los Angeles: University of California, Graduate School of Architecture and Urban Planning. (unpublished)

——(1982) "Technology, the labor process, and the location of industries." Ph. D. dissertation, University of California, Berkeley.

SYLOS-LABINI, P. (1969) Oligopoly and Technical Progress. Cambridge, MA: Harvard University Press.

THOMAS, M. and R. LETTERON (1975) "Perspectives on technological change and the process of diffusion in the manufacturing sector." Economic Geography 51: 231-251.

VERNON, R. (1979) "The product cycle hypothesis in a new international environment." Oxford Bulletin of Economics and Statistics 41(4): 255-267.

——(1966) "International investment and international trade in the product cycle." Quarterly Journal of Economics 80(2): 190-207.

WALKER, R. and M. STORPER (1981) "Capital and industrial location." Progress in Human Geography 5(4): 473-509.

WEBER, A. (1909) Uber den Standort der Industrien. Tubingen; J. C. B. Mohr.

13

Studying Technology and Social Life

CLAUDE S. FISCHER

□ THE BITS AND BYTES and discs and digits of the new communications systems are filled with speculations about what these new media might do to our lives. But well-documented and believable claims are hard to find. That is also true when we ask about the consequences of earlier communications technologies. Knowing how, say, the telephone affected social life might help us understand where microcomputers and optical fibers are taking us. Yet we know remarkably little about the social consequences of that technology—indeed, about the effects of many significant technologies.

I will describe here the sorry state of the sociology of technology—examine its decline, theoretical confusion, and empirical vacuum—and argue for new efforts to understand how technology influences social life. And I will venture some tentative ideas about how we might profitably study the social significance of technology.

The substantive concern here is specifically with technology's role in daily personal and social life rather than with large economic and institutional domains and with changes across one or two generations rather than across epochs of history. And I will draw most of the illustrations from my current research on the social role of the automobile and telephone in early twentieth century United States.

AUTHOR'S NOTE: *This article was supported in part by the National Science Foundation (Grant SES83-09301) and the National Endowment for the Humanities (Grant RO-20612). It was improved by comments from Melanie Archer, John Chan, Stephen Derné, Barbara Loomis, Ann Swidler, and Joel Tarr, but the remaining errors are solely my responsibility.*

A LOST TRADITION

The study of technology, especially of technologies used to transcend space, has a notable history in American sociology. Its theoretical concern, imported from Europe by the Chicago school, was how modernization altered the "natural" and "moral" orders of the community. Conditions of modern urban life, it was argued, sundered social relations and undermined social norms. New transportation and communications media were major elements of that modernization.

This concern generated studies of how technological changes, via economic, demographic, and organizational changes, altered the spatial form of the city—work identified with Burgess and continued by scholars such as McKenzie, Hawley, and Kasarda. In addition, sociologists investigating community change in the 1920s and 1930s looked closely at the role of transportation and communications in local social integration (see, for example, McKenzie, 1968; McClenahan, 1929; Coale, 1924; Brunner and Lorge, 1937).

William F. Ogburn was the leading figure in this investigation. The Chicago sociologist and his associates explicitly analyzed the contexts and consequences of technological change. Their works include essays on the sources of invention, the effects of technologies on the family, and projections of the social ramifications of aviation. Ogburn authored the major theory in the field, "culture lag," arguing that, as a general rule, there is an extended delay between the applications of a new technology and social adjustments to those applications. The interim is rife with dislocations and problems.

Work along these lines proceeded into the 1950s and then apparently stopped. As Westrum (1983:1) has put it, this flourishing and important enterprise "passed into oblivion in slightly more than two decades." No third generation of Ogburn sociologists has carried on the program; few articles on technology appear in the major sociology journals. Historians and futurologists now dominate its study and, despite widespread public interest in technology, the number of social science papers on it has dropped since the early 1970s (Westrum, 1983).

Whatever the reasons for this decline,[1] there is relatively little scholarship today on the social consequences of technology. One can read a great deal about the influence of society on technology—how academic settings, business interests, cultural perspectives, values, and the like shape the creation and use of technologies. But there is surprisingly little research literature on the other side of the relationship.

To be sure, there continues to be lively work on how new technologies alter production and commerce—a long tradition in industrial and agricultural economics. And sociologists, today many with a Marxist orientation, continue to study how the tools of labor affect workers. One thinks, for example, of the research on deskilling stimulated by Braverman (1974) and of the current rush to study computers in the workplace. But on how technology affects social relations, community, lifestyles, and what the French historians have called *mentalité*, American scholars have produced only fragments.

Into the scholarly vacuum on the personal consequences of technology have stepped the popular essayists, the Tofflers, Elluls, and Packards. Their motivating concerns—what are machines, electronics, and their ilk doing to our lives?—justifiably capture a wide audience. And their dramatic metaphors—shocks and waves, global villages, and wired societies—even pervade academics' discourse. Indeed, many academic theories of modern society rest on a set of largely unexamined assumptions about technology not unlike those of the popular essayists.[2]

Until there is a reconstructed sociology of technology, this seems to be our lot. But to rebuild and rise beyond the Ogburn tradition will require some difficult theoretical and empirical work.

SOME PROBLEMS IN CURRENT THINKING ABOUT TECHNOLOGY AND SOCIETY[3]

Certain intellectual habits impede scholarly and lay thinking about technology. Let us deal with them explicitly (see also Winner, 1977). One habit is interpretation through sweeping metaphors. Many writers use overarching concepts to analyze technological change—concepts such as rationalization, privatization, technique, and control—and they use metaphors to make vivid their interpretations—metaphors such as the machine, the clock, and the assembly line. What technology does, according to these interpretations, is all-pervasive and it parallels the metaphor. Modern life as assembly line is a powerful image, of course, as vividly demonstrated by Charlie Chaplin.

But there are subtler examples, as well. In Boorstin's *The Americans: The Democratic Experience* (1973), modern American life is shaped by dozens of technological innovations, from paper bags to the automobile. The key is "attenuation." All these technologies, Boorstin writes, "enlarged the daily experience of millions; but spreading also meant thinning. *Attenuation* summed up the new quality of experience" (Boorstin, 1973: 306). "Americans were increasingly held together not by a few iron bonds, but by countless gossamer webs knitting together the trivia of their lives" (148). Elsewhere, Boorstin (1965: viii) discusses the automobile as a bitch goddess forever demanding sacrifice. Other popular writers, such as Mumford and Ellul, generalize in similar ways. And, I would contend, this style of thinking pervades more scholarly works as well, in which researchers may appreciate complexity in the subjects immediately under their microscopes but still broadly label everything around it in global terms.

One problem with these sweeping metaphors is that they break up under close examination. Take, for example, the clock metaphor: Technology increasingly regiments and regulates people's use of time (from preindustrial to industrial time). Some modern technologies do, but some do not. The railroad, for example, probably encouraged regimentation by requiring people to meet its schedules; it even stimulated standardized time zones. But the automobile loosened the scheduling, as drivers were able to come and go as they pleased.[4]

Recently, Kern (1983) has claimed that around the turn of the century the bicycle and automobile helped form a new *zeitgeist* of speed. No doubt, riding these machines gives one a sensation of speed. But, on the other hand, these new vehicles could just as well have contributed to a greater sense of leisureliness. Farmers, for example, could now sleep later on the mornings they went to market and dawdle longer in town before returning home.

Yet another example of the confusion wrought by sweeping generalizations is the common tendency to confuse bureaucracy with technology (Winner, 1977: chap. 5). Each—tools and organizations—involves rationality; each elicits the cog-in-the-machine metaphor. Yet brief reflection suggests that the two should be separated. Huge bureaucracies, such as feudal manors, the Imperial Chinese administration, and the Incan Empire, were organized with very little technology.

Perhaps ultimately and abstractly, sweeping together such disparate elements as screwdrivers, LOGO, and seniority rules under labels such as "technique" may be valid, but it seems to hinder any serious study of technological effects, to obscure more than it reveals.

Similarly, many discussions of technology assume homogeneous effects: that all technologies operate in the same direction. Mumford, for example, states:

> During the last two centuries, a power-centered technics has taken command. . . . By now, a large part of the population of the planet

feels uneasy, indeed deprived and neglected unless it is securely tied to the megamachine: to an assembly line, a conveyor belt, a motor car, a radio or a television station, a computer or a space capsule. . . . [E]very autonomous activity . . . has either been bulldozed out of existence or reshaped. . . . to conform to the requirements of the machine" (1972: 4-5).

In the same vein, many commentators loosely assume, for example, that modern communications systems reduce individual autonomy. But on closer examination, one can argue so for broadcast media (such as television), but not easily so for point-to-point media (such as the telephone), which probably increase individual autonomy.

A Chicago school dissertation on the social aspects of the automobile in the late 1920s illustrates the conceptual tropism toward homogenizing all technologies. In early chapters Mueller (1928) presents evidence on the complex community role of the automobile. One chapter, for instance, makes a persuasive case that the automobile helped farmers organize, fraternize, and form communities. Nevertheless, Mueller concludes that the car is yet one more innovation bringing anomie, transitory relations, and the like.

The homogeneity assumption appears in the literature in other ways, as well: in implicit claims that all the consequences of a specific technology operate in parallel (e.g., that the automobile destroys community by fostering detached housing, and by isolating drivers on the road, and by creating status competition, *and* so on) or that all people are affected in same ways—rich and poor, urban and rural, men and women.

But there is no a priori reason that different technologies, or even similar ones, need have congruent effects, nor that all their effects need to be consistent, nor that everyone or every society need react similarly. Perhaps, for instance, the automobile fostered social cohesion in rural areas but undermined it in cities, or benefited blacks more than whites (Preston, 1979).

A third and related habit is to assume that the consequences of a technology are linear, that its effects cumulate in the same direction. Effects could, instead, reverse. For example, transportation improvements probably affected recreational travel in rural America in contradictory ways. Touring and staying overnight at rural taverns, common in the horse-stage era, declined when the railroad encouraged traffic between towns and discouraged use of the local, capillary road systems. But then the automobile may have restimulated countryside travel and inns, partly by helping stimulate the rebuilding of those local roads (Berger, 1979: 120-121; Belascoe, 1979).

These reversals may even occur in the history of a single technology. The Lynds suggest in *Middletown* (1929: 174-175) that the first washing machines stimulated collective housework in laundromats, but as the machines became cheaper, women once again did their laundry at home (see also Strasser, 1982). Similarly, Pool (1980) argues that the telephone first contributed to urban centralization and then to urban decentralization.[5] Detailed examination of specific technologies may reveal more complex and time-dependent effects than many current commentaries imply.

A fourth somewhat different habit might be labeled the impact-imprint assumption, the notion that a technology "impacts" society by transfering its own properties to individuals, groups, or institutions; that there is isomorphism between forms and contents of cause and effect. (The use of metaphors noted above helps to sustain this mode of thinking). One sees this, for example, in the assumption that labor-saving devices must lead to less labor in the home; or that the use of a machine for talking (the telephone) makes conversation impersonal; or that the mobility of the automobile causes people to be mobile; or that space-transcending technologies in general cause people to lead space-transcending lives.

Just as the basic causal assumption need not be so (see Fischer, 1970: 177-178), the specific applications need not be so. The latest research suggests that the time spent on housework has not declined, despite time-saving appliances (Vanek, 1978); that the telephone is used to sustain very personal relations (Singer, 1981; Synge et al., 1982); that residential mobility has actually declined in the twentieth century (Long, 1976), perhaps in part because the automobile allows people to stay in the same residence even when they change jobs; and that transportation and communications media may have encouraged more local than distant interactions. On the last point, Willey and Rice (1933) argue well that automobiles and telephones stimulated far more local contacts than they did distant ones. Instead of breaking down "island communities" (Wiebe, 1975), these space-transcending technologies may have reinforced them. In these sorts of ways, the impact-imprint logic is pervasive and pervasively misleading.

As an aside, I note a complementary logic: reaction or compensation. People react against or compensate for the properties of a new technology. For example, mass media transmit many messages; therefore, people compensate by, say, psychologically withdrawing. Or, return-to-nature movements are "natural reactions" to an increasingly machine-driven world. This process may sometimes happen, but it is too facile an assumption to be followed unreflectively, as it often is.

Beyond these conceptual difficulties is a larger problem: that most discussions of technology are based on commonsensical extrapolations rather than empirical research. Indeed, there is an empirical vacuum of startling proportions on the social consequences of technology. There are few major works. One thinks of Pred's (1973) analysis of how the telegraph altered the geography of America's economy and of Weber's (1976) account of how improved roads helped end the isolation of French peasants; but these are rare works. And outside of a few very specific topics—for example, the mechanical loom, the printing press, television—there appears to be no cumulation of empirical research.

One promising line of study concerns household technologies. Although of mixed quality and bearing little sociological input, recent histories of housework suggest how new tools might have altered women's and families' daily lives (see, for example, Cowan, 1976; Strasser, 1982; Vanek, 1978). Many have written on the effects of mass media, although these essays tend to be either narrowly focused empirical studies, such as those that social psychologists have done on television and aggression, or blunder-buss polemics by culture critics. Some serious but isolated work has been done on the effects of fertility-control technology and demographic change.

There have been efforts outside of sociology to address social consequences. The National Science Foundation briefly supported "retrospective technology assessments" of a few major innovations, such as the underseas cable, with hopes of developing forecasting tools (see Tarr, 1977; Coates and Finn, 1979; Pool, 1983). The products have been, in my estimation, thoughtful but relatively speculative and commonsensical extrapolations, rather than either substantial theories or well-documented studies, at least when the authors have ventured to look at social and cultural consequences. Similarly, it is my impression that the various "technology and society" programs around the country, by necessity, stress history of technology and policy debates about technology, while relying on speculative material for covering social consequences.

Empirical knowledge about the social consequences of technology is slim. Below, I will illustrate this claim further with an obviously significant technology, the automobile. The point here is that even clear, logical thinking, in the absence of solid evidence, is likely to go wrong.[6]

For these and, no doubt, other reasons, current writing on the social consequences of technology is too often unsatisfying and unexciting, even though the subject is patently important and popular. In the next section I will pursue the theoretical and empirical difficulties found specifically in the study of the automobile and telephone.

LUCANAE IN THE STUDY OF
THE AUTOMOBILE AND TELEPHONE

We know much about the history of automobile production and its role in the economy. We also have studies of how the automobile may have shaped metropolitan form.[7] But what do we have, first, in the way of a theory of the automobile's social role?

Not much. Automotive historians' accounts of social effects tend to be ad hoc extrapolations from manifest uses of the automobile. This includes Ogburn's associates as well (Allen, 1957; see also Rae, 1965). Some scholars have loosely applied general ideas about modernization and anomie to the automobile (e.g., Mueller, 1928; McClenahan, 1929). Others have focused on the automobile as icon (e.g., Brownell, 1976) or as a costly consumption good (Flink, 1976; Lynd and Lynd, 1929).

The closest to a general theory of the automobile and other space-transcending technologies comes from geographers' writings on the "friction" of space. The cost and time of moving across space condition all action; distances, therefore, significantly influence resulting social patterns. New technologies reduce that friction of distance, thereby altering habits and institutions conditioned on it. In other words, such technologies create a "geography" where all places converge. One implication is that space-transcending technologies undercut the rationale for cities, proximity, and for selecting locations according to access, because all places increasingly have equal and immediate access to one another (see Abler, 1975; also, Abler and Falk, 1981; Falk and Abler, 1980; Webber, 1964; Gottman, 1977).

Although useful, this model is far from complete for tracing social consequences. It does not tell us whether people who save time and financial expense in travel spend those savings in locating further away and holding constant travel expense (which might have been the automobile's effect on the separation of home and workplace), in making more frequent trips at the same total expense (which may have happened in rural neighborhoods), or in some other fashion, such as leisure. It does not tell us whether the technology led to trips never before contemplated, such as camping, or to an entirely new system of destinations, such as the decentralized retail system of contemporary metropolises. Nor does it begin to deal with the social consequences of the technology outside of its specific use—its symbolism, role in the household budget, cultural role, and so forth.

Ithiel de Sola Pool (1977: 4) notes the analytical complexities when he writes about the telephone, but it fits the automobile too:

> Its impacts are puzzling, evasive, and hard to pin down. . . . [It] adds to human freedom, but those who gain freedom can use it however they choose. Rather than constraining action in any one direction, [it] is an agent of effective action in many directions.

> That conclusion does not imply that the phone has no impact or that there is nothing to study. On the contrary, it implies that the study of the telephone's social impact belongs to the important and subtle class of problems in the social sciences which demands a logic more complex than that of simple causality—a logic that allows for purposive behavior as an element in the analysis.

That kind of comment appears frequently in dispassionate analyses of the telephone (those not written by industry publicists). "It may be that the social impact of the telephone is so difficult to assess because it is such an adaptable and unobtrusive tool, the use of which is molded by individuals and society to the pursuit of diverse aims" (Gottman, 1977: 316). And, indeed, there is little theory of telephony per se.

Aronson (1971) has suggested that telephone use facilitated large-scale organization, stimulated retail trade, created "psychological neighborhoods" of physically dispersed social relations, sustained the extended family, provided a sense of security, and stimulated rural sociability. Ball (1968) has discussed subtler concomitants of the telephone, particularly the ways that self-presentation and interaction management differ between telephonic and face-to-face conversations. Mostly, we have many ad hoc claims—dozens of early ones are catalogued by Pool (1983)—and several informed speculations—most in a volume edited by Pool (1977)—but little comprehensive theorizing that is not derived from the general friction model just described.

Perhaps theories of the automobile and telephone should be induced from empirical materials. Unfortunately, the empirical literature on these two technologies also is slim.

There is little strong evidence on the automobile, considered by many to be the most consequential technology of the century. To be sure, many scholars, Ogburn and his associates among them, have discussed the automobile's influence on community, family, religion, morality, art, and the like (see, for example, Lewis, 1980; Flink, 1980). But their actual evidence concerns topics such as automobile sales, mileage driven, and direct economic correlates, such as the demise of the horse—not social effects (e.g., Allen, 1957). Their comments on the latter are largely informed speculations. Historians have contributed some excellent accounts of the automobile's diffusion and reception (e.g., Flink, 1970; Wik, 1972; Berger, 1979; Preston, 1979). Although very useful, these studies essentially review contemporary public opinion about the automobile's social effects—a necessary beginning, but insufficient for drawing conclusions about the effects themselves.

There is one set of findings that seems to inform all historians' and sociologists' discussions of social effects: the chapter in *Middletown* that begins with a resident's comment:

> Why on earth do you need to study what's changing this country? I can tell you what's happening in just four letters: A-U-T-O! (Lynd and Lynd, 1929: 251).

Again and again, the Lynds' observations, quotations, and interviews are the key or only source for assertions about social effects such as tension in the family, decline in the church, and increased illegitimacy (e.g., see Flink, 1976: 140-160). Whatever the evidentiary value of the Lynds' materials—much of it is impressionistic and retrospective—alone it is a weak reed for so much to rest on.[8]

Instead, we often have contradictory empirical claims: that the automobile encouraged church attendance, at least among isolated countryfolk, or discouraged it by providing alternative Sunday leisure; that it strengthened the family through joint trips, or broke it up by allowing separate trips; that it undermined rural cohesion by allowing farmers to find sociability far away, or that it strengthened the rural neighborhood by permitting frequent visits and gatherings; and so on.

On the telephone we have even less: histories of the industry (particularly of AT&T) and of the hardware, but essentially nothing on the social side, excepting some contemporary social psychological studies trying to contrast the use of telephonic versus face-to-face contacts. Pool's collection, *The Social Impact of the Telephone* (1977), contains the best of informed speculations, but not much evidence. On this topic, too, we often have contradictory claims: that the telephone encouraged farmers to stay on the farm, or to leave the farm; made people more, even too, sociable, or made them less sociable; created a sense of security, or a "habit of tenseness" (Brooks, 1976: 117); or just unexamined assertions. (My favorite example appears often in the slim telephone literature and probably was first presented by AT&T's chief engineer in 1902, J.J. Carty:

that the telephone made possible the skyscraper, because without it, all the interior space of the buildings would have to be filled with elevators carrying messenger boys from floor to floor. Why? It seems obvious that pneumatic tubes could have been used.)[9]

That these sorts of unsystematized fragments of research, or more properly, suggestions for research, typify the literature on two of the most pervasive technologies of the twentieth century underlines the generally impoverished state of research on personal technology.

AN AGENDA FOR A NEW SOCIOLOGY OF TECHNOLOGY

Rehabilitating the sociology of technology will call for both more theoretical sophistication and more detailed empirical work. I have no "theory of technology" to offer. Indeed, I have argued above that homogenizing all technologies into one "techne" is likely to be an error. But there are a few metatheoretical guides that may be useful. At least, they seem useful in my current research on early twentieth century communications and transportation.

One is, of course, to avoid the analytical problems discussed earlier: sweeping metaphors, assumptions of homogeneity and of linearity, and the implicit use of the impact-imprint model. Beyond that, I would argue for specificity. Until persuasive argument or, better yet, solid evidence shows otherwise, we should assume that each technology is used differently and has distinct consequences and, moreover, that the same technology may be used differently by different people in different ways to different effect. This calls for close-grained examination of specific tools and techniques.

Our models should clearly incorporate agency. Few would argue, upon reflection, that a technology really forces, pushes, or impels people to act in certain ways. But the understanding that people choose to use a technology or not or how to use it tends to get trampled in the rush to generalize. Agency is stressed in studies of how technologies are created; we now appreciate that new systems do not emerge from an autonomous line of invention, but that people with money and power make decisions about investment, marketing, and the like, and that historical circumstances stall or detour innovation.[10] The same awareness needs to be incorporated into analyzing the consequences of those technologies. We cannot assume that people use a technology because "it is there" or because it is "obviously" advantageous, nor because they have been "brainwashed" to buy it, nor because they have been swept along by a cultural ethos. Nor will they necessarily use it as it was designed to be used.[11] We need to ask how and why purposeful actors choose to adopt specific technologies and what they do with them.

This leads us to think of technologies as facilitators of human action, rather than as directors of it. Pool (1983) makes this point about the telephone, but it has much wider applicability. We must understand what it is that people are trying to do for which new technologies are new means. How do these technologies help people attain their ends in new ways? For example, how did the automobile facilitate marketing, shopping, visiting, and courting among rural Americans? It may be that people continued to do what they always had done, only more cheaply and comfortably; that, for example, people visited their in-laws as often as before automobiles, but did so more cheaply now, or that promiscuity was no less common, but that it now happened in a place softer than a hayloft. If that is the answer to our studies, it is no less interesting than a conclusion that social life was dramatically altered.

This should not imply that new technologies are no more than means for continuing existing patterns. That is one skeptical approach to understanding technology (see Daniels, 1970). But there must be more (although not necessarily as much more as some assume). The existence and use of a technology alters material and social givens, creating new options for and new constraints on individual action. ("Structuration" is, I think, the current term for this process.) The automobile's widespread use may have brought into being, through better roads and roadside accommidations, a new possibility of touring by car. Or, when the automobile became commonplace, perhaps new expectations arose for people to be places—say, to pick up their children at school—they had not previously been expected to be. In this manner, the technology becomes more than a tool for action; it becomes a condition of action.[12]

Conceptually, then, it seems that we want to ask two pairs of questions, in sequence: First, why and how do people use a technology to pursue their ends? And how does using that technology lead them to alter other aspects of their lives? For example, perhaps many people obtained and used the telephone largely for keeping in touch with their immediate family. That, in turn, may have led many to reduce the frequency of personal visits to kin and to use that time instead for sociability with nonkin.

Second, how does the common use of a technology alter the options and constraints for action? And how do the resulting patterns of action create further structural and cultural circumstances? For example, common use of the automobile may have accelerated suburban housing construction by expanding the geographical pool of homebuyers. That, in turn, may have diverted so many resources from center-city housing so as to foreclose the latter as an option for many (and, thus, reinforce the suburban demand).

This approach explicitly incorporates agency and conditions of action in examining technology's role. And the structured form of the questions serve us better, I think, than the simple "billiard-ball" model of primary, secondary, and tertiary effects left us by Ogburn. Still, of course, there are many ambiguities left. One involves drawing bounds. Few would be so bold or ambitious as to trace, for instance, how the medieval chimney over centuries created modern individualism (White, 1974: 8-9). We have much work to do even with closer horizons.

These are questions I have presented, not theories. And we would do well to build any theories in conjunction with empirical research, given the paucity of facts. Theory and data are grist for one another, but I fear we have more stimulating hunches than sound data in the study of technological consequences. So my major call is for full, detailed, painstaking study of specific technologies and their consequences, bringing to bear the widest array of social scientific techniques, from analyzing personal diaries to see how people confronted and applied a technology to time-series statistical analyses to see what patterns of action changed in step with the diffusion of a technology. Such empirical work may show us much more detailed connections between technology and society, more subtle interactions, and perhaps less cataclysmic changes than much current commentary would suggest.

Cataclysmic, indeed, is the way some contemporary observers describe the coming effects of computers and other electronic marvels. We are in a new age, a third wave, a megatrend, an "informatic" postindustrial turning point in history. Perhaps. But the evidence so far has not been marshalled for these claims—indeed, may not be possible to marshall for years. Already, forecasts of a wired society seem premature. Projections of how many computers will invade the home are falling short (Sanger, 1984: 1), in part, because analysts are not sure why people would want home computers in the first place.

The newest information media are certainly significant. But they will call for close, specific, even mundane empirical examination before we can project social consequences: What do people do with these tools? For example, do most people just play games on their computers? What do people do because they save or spend more time, energy, and money by using these machines? Do the game players now watch less television? What would commonplace ownership of the new machines mean for others? For instance, if most households had telephone answering machines with call-in features, would a cultural expectation arise that everyone should be able to receive a message within a few hours, no matter where they were? Would that, in turn, perhaps lead to social inventions to replace the "Sorry, I missed your call" excuse? And so on.

Writing these words on a computer, I can hardly dismiss all the speculations about the marvels (or terrors) of the silicon-chipped society. But if we have any historical and sociological sense—if we avoid what White (1974: 3) calls the "Hudson Institute syndrome"—we should remain skeptical about major social changes until the evidence is in.

NOTES

1. Merrill (1968: 576) notes that the study of technology lacks major institutional support in academia and suggests that one reason is that "technologies are not thought to be very interesting" compared to, say, the intellectual challenges in the sociology of science or knowledge. Westrum (1983) attributes the decline in part to counter-cultural attacks on technology during the 1960s and in part to the ascendance of historians of technology who emphasize particularly over sociologists' generalizations. Or perhaps historians have ascended because the truly major technological changes of our epoch (excepting television), such as the automobile, occurred before World War II. One could also argue that the sociology of technology was never well-endowed with powerful theories. Other than notions about cultural lag and simultaneity in inventions, there was little theory to drive research. And, on the other side, much of the empirical work seems descriptive and commonsensical and not likely to stimulate much interesting theory.

2. For example, some modernization theorists present ideal types that assume strong links between communications technologies and social qualities such as impersonality and bureaucracy; modern society is defined in terms of these undocumented causal connections.

3. An earlier version of these comments was presented to a colloquium at the Department of Sociology, University of Chicago, October, 1982.

4. Similarly, television programming encouraged simultaneous viewing, but the video recorder may now be freeing viewers from network schedules.

5. It encouraged centralization up to about 1910 by permitting businesses to concentrate downtown while staying in touch with suppliers, customers, and branch offices over the telephone. Afterwards, the telephone, together with the automobile, encouraged decentralization by making all metropolitan locations almost equally accessible (Pool, 1980).

6. It would seem that the introduction of advanced technologies into developing nations would provide a wealth of research opportunities for understanding their social ramifications. Yet these opportunities—for example, the provision of modern telecommunications—seem not to have been exploited in any systematic or rigorous fashion to inform the sociology of technology.

7. See Borchert (1967) and Hawley (1981) for overviews and Guest (1972) for an illustrative study. Some researchers have claimed, however, that the automobile's role in urban dispersion has been overestimated (e.g., Foster, 1975).

8. D.N. Pappenfort's (1964) monograph is an interesting but lonely direct empirical study of the automobile's social effects. He presents persuasive data that the automobile allowed more rural mothers to have their children in hospitals.

9. The other part of Carty's claim is that constructing the buildings required telephonic communications between the ground and high floors. Again, I see no reason that semaphore systems, long in use, could not have sufficed.

10. See, for example, the accounts of national variations in the implementation of the telephone presented in Pool (1977).

11. As illustration, consider recent research indicating that people frequently use television as a background to other activities, including conversation, rather than as a direct object of attention. On the gap between how marketers of the telephone saw their product and how customers employed it, see Fischer (1983).

12. I allude here, of course, to Marx's dictum about men making their history, but not under conditions of their own choosing.

REFERENCES

ABLER, R. (1975) "Effects of space-adjusting technologies on the human geography of the future," pp. 35-56 in R. Abler et al. (eds.) Human Geography in a Shrinking World. Belmont, CA: Duxbury.

———and T. FALK (1981) "Public information services and the changing role of distance in human affairs." Economic Geography 57(1): 10-22.

ALLEN, F. R. (1957) "The automobile," pp. 107-132 in F. R. Allen et al. (eds) Technology and Social Change. New York: Appleton-Century-Crofts.

ARONSON, S. H. (1971) "The sociology of the telephone." International Journal of Comparative Sociology 12(September): 154-158.

BALL, D. W. (1968) "Toward a sociology of telephones and telephoners," pp. 59-75 in M. Truzzi (ed.) Sociology and Everyday Life. Englewood Cliffs, NJ: Prentice-Hall.

BELASCOE, W. J. (1979) Americans on the Road: From Autocamp to Motel, 1910-1945. Cambridge, MA: MIT Press.

BERGER, M. L. (1979) The Devil Wagon in God's Country: The Automobile and Social Change in Rural America, 1893-1929. Hamden, CT: Archon Books.

BOORSTIN, D. J. (1973) The Americans: The Democratic Experience. New York: Vintage.

———(1965) "Editor's preface," pp. vii-ix in J. B. Rae, The American Automobile. Chicago: University of Chicago Press.

BORCHERT, J. R. (1967) "American metropolitan evolution." Geographical Review 57(July): 301-332.

BRAVERMAN, H. (1974) Labor and Monopoly Capital. New York: Monthly Review Press.

BROOKS, J. (1976) Telephone: The First Hundred Years. New York: Harper & Row.

BROWNELL, B. A. (1976) "A symbol of modernity: attitudes toward the automobile in southern cities in the 1920's." American Quarterly 24(1): 20-44.

BRUNNER, E. de S. and I. LORGE (1937) Rural Trends in Depression Years: A Survey of Village-Centered Agricultural Communities. New York: Columbia University Press.

COALE, J.J. (1924) "The influence of the automobile on the city church." The Annals 116(November): 30-82.

COATES: V.T. and B. FINN (1979) A Retrospective Technology Assessment: Submarine Telegraphy. The Transatlantic Cable of 1866. San Francisco: San Francisco Press.

COWAN, R.S. (1976) "The 'industrial revolution' in the home: household technology and social change in the 20th century." Technology and Culture 17(1): 1-23.

DANIELS, G.H. (1970) "The big questions in the history of American technology." Technology and Culture 11(January): 1-21.

FALK, T. and R. ABLER (1980) "Intercommunications, distance, and geographical theory." Geografiska Annaler 62B(2): 59-67.

FISCHER, C.S. (1983) " 'Educating the public': selling telephones and automobiles in the early days." Presented to the Social Science History Association, Washington.

FISCHER, D.H. (1970) Historians' Fallacies. New York: Harper & Row.

FLINK, J.J. (1980) "The Car Culture revisited: some comments on the recent historiography of automotive history." Michigan Quarterly Review 19(4)-20(2): 772-781.

———(1976) The Car Culture. Cambridge, MA: MIT Press.

———(1970) America Adopts the Automobile, 1895-1910. Cambridge, MA: MIT Press.

FOSTER, J.S. (1975) "The Model-T, the hard sell, and Los Angeles' urban growth: decentralization of Los Angeles during the 1920's." Pacific Historical Review 44(November): 459-484.

GOTTMAN, J. (1977) "Metropolis and antipolis: the telephone and the structure of the city," pp. 303-317 in I. de Sola Pool (ed.) The Social Impact of the Telephone. Cambridge, MA: MIT Press.

GUEST, A. (1972) "Urban history, population densities, and higher status residential location." Economic Geography 48: 375-387.

HAWLEY: A. (1981) Urban Society. New York: John Wiley.

KERN, S. (1983) The Culture of Time and Space, 1880-1918. Cambridge, MA: Harvard University Press.

LEWIS, D.L. [ed.] (1980) "The automobile and American culture." Michigan Quarterly Review 19-20 (Fall): entire issue.

LONG, L.H. (1976) "The geographical mobility of Americans," in Current Population Reports. Special Studies, Ser. P-23, 64. Washington, DC: U.S. Bureau of the Census.

LYND, R.S. and H.M. LYND (1929) Middletown. New York: Harcourt Brace Jovanovich.

McCLENAHAN, B.A. (1929) The Changing Urban Neighborhood: From Neighbor to Nigh-Dweller. Los Angeles: University of Southern California.

McKENZIE, R. (1968) "The neighborhood," pp. 51-93 in A. Hawley (ed.) Roderick D. McKenzie on Human Ecology. Chicago: University of Chicago Press. (originally published, 1921)

MERRILL, R.S. (1968) "The study of technology," pp. 577-589 in International Encyclopedia of the Social Sciences, vol. 15. New York: Macmillan.

MUELLER, J. H. (1928) "The automobile: a sociological study." Ph. D. dissertation, University of Chicago.

MUMFORD, L. (1972) "Two views on technology and man," pp. 1-16 in C. A. Thrall and J..M. Starr (eds.) Technology, Power, and Social Change. Carbondale: Southern Illinois University Press.

PAPPENFORT, D. N. (1964) Journey to Labor: A Study of Births in Hospitals and Technology. Chicago: University of Chicago, Population Research and Training Center.

POOL, I. de S. (1983) Forecasting the Telephone. Norwood, NJ: Ablex.

———(1980) "Communications technology and land use." The Annals 451(September): 1-12.

———[ed.] (1977) The Social Impact of the Telephone. Cambridge, MA: MIT Press.

PRED, A. R. (1973) Urban Growth and the Circulation of Information: The United States' System of Cities, 1790-1840. Cambridge, MA: Harvard University Press.

PRESTON, H. L. (1979) Automobile Age Atlanta. Athens: University of Georgia Press.

RAE, J. B. (1965) The American Automobile: A Brief History. Chicago: University of Chicago Press.

SANGER, D. E. (1984) "The expected boom in home computers fails to materialize." New York Times (June 4): 1.

SINGER, B. D. (1981) Social Functions of the Telephone. Palo Alto, CA: R&E Research Associates.

STRASSER, S. (1982) Never Done: A History of American Housework. New York: Pantheon.

SYNGE, J., C. J. ROSENTHAL, and V. W. MARSHALL (1982) "Honing and writing as a means of keeping in touch in the family of later life." Presented to the Canadian Association on Gerontology, Toronto.

TARR, J. A. [ed.] (1977) Retrospective Technology Assessment—1976. San Francisco: San Francisco Press.

VANEK, J. (1978) "Household technology and social status." Technology and Culture 19(July): 361-375.

WEBBER, M. M. (1964) "The urban place and the nonplace urban realm," pp. 79-153 in M. M. Webber et al. (eds.) Explorations into Urban Structure. Philadelphia: University of Pennsylvania Press.

WEBER, E. (1976) Peasants into Frenchmen: The Modernization of Rural France, 1870-1914. Stanford, CA: Stanford University Press.

WESTRUM, R. (1983) "What happened to the old sociology of technology?" Presented to the Society for the Sociological Study of Science, Blacksburg, VA.

WHITE, L., Jr. (1974) "Technology assessment from the stance of a medieval historian." American Historical Review 79(February): 1-13.

WIEBE, R. (1975) The Segmented Society: An Introduction to the Meaning of America. New York: Oxford University Press.

WIK, R. M. (1972) Henry Ford and Grass-Roots America. Ann Arbor: University of Michigan Press.

WILLEY, M. M. and S. A. RICE (1933) Communication Agencies and Social Life. New York: McGraw-Hill.

WINNER, L. (1977) Autonomous Technology: Techniques-out-of-Control as a Theme in Political Thought. Cambridge, MA: MIT Press.

Part VI

Alternatives

14

Which "New Technology"?

DOREEN MASSEY

□ OVER THE LAST FOUR DECADES there have been, as in many other countries, major transformations in occupational and spatial structure within the United Kingdom. Much of what has disappeared has declined, not as a product just of short- or medium-term recession, but as a product of long-term structural change. It is unlikely to reappear or be rebuilt. The future is likely to be very different from the past. The changes in occupational and spatial structure reflect deeper underlying shifts—technological change being one of them—and, perhaps more important in the context of the discussion here, are likely in turn to be the basis for much more far-reaching social changes.

These changes are already occurring; indeed, they are well under way. There is agreement that somewhere in the complex hub of causal relations that lie behind them, technological change plays a significant role. The attitude of politicians, of academics, and of the public at large through the influence of the media is worth reflecting upon. When questions of the future, and of the future implications, of these processes are addressed, discussion usually takes the form of prediction of the future rate and nature of technological change, evaluation of its likely social effects, and recommendations as to how these effects, which are often viewed as not-entirely-positive, might be mitigated.

Prediction of technology and evaluation of its social implications: The difference is striking. It puts technologists, scientists, and the present structure within which they operate (corporate, military, state, etc.) in the position of unchallenged originators of the new future—something like modern-day surrealists whose role in the process of creation is merely to be the medium through whom the message passes. Social scientists, on the other hand, having waited for these pronouncements on the shape of the new technology, and having seen something of its operation, are given the task of monitoring it and coping with its effects. Technology can be predicted according to this formulation and society must, somehow or other, mold itself around it.

This is a formulation that reflects both the power relations within society and important aspects of the division of labor within academe, most notably, the separation of physical sciences and technology, on one hand, from the social sciences, on the other. It is a formulation, also, that increasingly is being challenged.

What I want to do in this chapter is explore some of the ways in which broad technological changes and, in particular, "new technology" have been finding expression within the British context, to explore some of their implications in terms of the nature of technological change itself, and then to look briefly at some of the alternatives that are beginning, still tentatively, to find their way onto the policy-making agenda.

THE BROAD SCENARIO

The main empirical focus of this chapter is on issues of occupational and spatial structure and it is helpful, to begin with, to lay out in the form of a broad sketch the general lines of the changes that have taken place.

At a national level, the United Kingdom has been experiencing many of the same kinds of changes in occupational and industrial structure that are occurring in other advanced capitalist industrial countries. Manual workers are declining as a proportion of the workforce; nonmanual workers are concomitantly increasing. Manufacturing employment has been falling for nearly 20 years, whereas employment in the service industries, over the longer term, continues to increase. The proportion of women in the (paid) labor force has increased markedly over the whole postwar period, although their absolute number has not grown so consistently, there having been a considerable absolute fall during the last half-decade of general economic devastation. Part-time work, which in the U.K. is penalized not only by lower pay but also by the absence of a wide range of legal benefits and protections, has increased in significance. As far as skills are concerned, within the generally expanding nonmanual groups it has been the administrators, managers, and professionals—the higher-status, higher-paid (and largely male) groups—that have been expanding fastest. The engineers and technicians group has both expanded and seen a substantial shift in its internal composition, in particular away from "old-fashioned" and more production-based engineering toward research and development scientists, the designers of software, and so forth. Clerical workers, too, have increased in importance as a proportion of the workforce. Among manual workers all categories have been declining.

The structural upheavals that lie behind these shifts are fundamental and long-term. They can by no means all be traced to technological change, still less to new technology. Perhaps more deeply than anything else, they mark a long-term reorientation of the British economy (although "toward what?" is a question still to be answered) and a shift in its place in the international division of labor. The reason for rehearsing these changes here is not that new technology has caused them (although it is certainly a contributor to some of them: the decline of craft jobs, the increase of R&D workers, for instance) but that the changes themselves form an important backcloth to understanding the social impact, in the widest sense, of "new technology." It is the context within which it is held up as the hope for the future.

NEW GEOGRAPHICAL PATTERNS

One important aspect of this wider context has been that over the last two decades there has also been a major change in the spatial organization of the British economy. In employment terms the old, basic sectors in the heavy industrial peripheral regions of the country have continued and usually accelerated their decline, leaving behind them high levels of male unemployment in areas in which once relatively highly paid manual jobs for men dominated the local economies. These have been long-term declines, although still continuing apace. In the 1960s and 1970s a, to some extent compensating, movement took place as jobs in a range of manufacturing industries (clothing, electronics assembly, light engineering) and, especially later, in services saw their most important growth in the less urbanized areas of the country and, also, in the old heavy industrial periphery. Overwhelmingly, such jobs were tedious, classified as low-skilled, certainly low-paid, and usually allocated to women. In total contrast, the same span of years (from the mid-1960s to the mid-1970s) saw the most significant increase in employment of managers and of the new breed of industrially related scientists and technologists. These jobs were taken by graduates, were classified as highly skilled, and were overwhelmingly allocated to men. Their increase derived, in the case of managers, primarily from the growing concentration of capital that took place in particular over this period and, in the case of scientists and technologists, from the changing balance in the technological base of the economy. The geography of this group was very different from that of the still-growing elements of production.

Managerial occupations increasingly clustered in and around the capital, increasingly separated, too, from the processes of production over which they exercised control. The new jobs in the new technology were also concentrated in the south and east of England, but this time less in London itself and more in parts of its semirural periphery, especially to the west along the M4 corridor to Bristol and around Cambridge to the north. Finally, while these patterns of new jobs were being established, the decline of jobs in manufacturing as a whole was gathering pace. And in recent years it has been this deindustrialization that has been the dominant element in the changing spatial organization of the British economy. For deindustrialization, too, has had its own specific geography. It has been the cities that have been hit hardest. From there, as the process gathered pace, deindustrialization has decimated the economies of the old manufacturing regions, in particular the west midlands and the northwest of England. Since the late 1970s even the new jobs decentralized to "the regions" only a few years before have begun to disappear (Massey, 1984).

As was stated above, these changes are, to a large extent, the outcome of major, long-term structural shifts, many of which had their origin at an international level. The British economy is clearly slipping down the ranks as a major producer and exporter of manufactured goods. Moreover, whereas in many of the industries that are now in decline the British economy was once internationally preeminent, this is less true of the new industries, particularly those based around high technology, which look likely to take over the role of economic pace setters. A distinction should be made here between British industry and the British economy, for the two are not coincident. British industry is performing far better than is the British economy, the latter being defined as the economic activity actually based within the geographical confines of the United Kingdom. This has implications for the nature of the emerging occupational and spatial structure to which I shall return in a moment. For the moment, though, what it is important to register is that these major structural shifts have left behind them urban and regional problems of a scale, an intensity, and a kind that means that they are quite simply insoluble by the old, well-tried, and all-too-frequently failed gamut of urban and regional policies. This implacable fact is now increasingly being recognized by a small but growing range of academics, by community groups, by workers in industry, and by local authorities.

"NEW TECHNOLOGY" AND URBAN AND REGIONAL PROBLEMS

In this context "new technology" is held up as either necessary to or responsible for either a bright new future or an ominous foreboding of mass unemployment. For some, new technology heralds the leisure society, removes the drudgery of work, provides a potential basis for regrowth for regions now in decline and dereliction. The technology belt of the M4 corridor and the Cambridge area is pointed to as a vision of the future for all. Others see nothing but unremitting gloom. New technology will lead not to leisure but to unemployment, not to more skilled jobs but to mindless or mind-numbing jobs, the bright new growth of parts of the southeast is contrasted with the low-paid tedium of the twilight shifts of the production end of the electronics industry in the peripheral regions of the country. All these scenarios, as they stand, take the shape of "new technology" as given. But for many of those who would foresee the more pessimistic future, it is precisely that which is now being challenged. The present section briefly outlines two different urban/regional situations in which this issue is now being put on the agenda. The alternatives that are being proposed will be examined in the last section of the chapter.

If one were to caricature the shift that has taken place in the economic and social geography of the United Kingdom between the last great crisis in the 1930s and that of today, one would point above all to the shift from a predominantly sectoral and urban pattern to one increasingly less dominated by the great urban concentrations and in which the prime differentiators between regions are no longer based on industrial structure (in the SIC sense) but on occupational structure, both within and between industries.

This new dimension of interregional inequality has become a major focus of debate within both academic and policy arenas. One of the fundamental questions at issue is how to characterize it. Descriptively, and as laid out in the earlier sections of this chapter, the issue appears to be one concerning the differential distribution of different types of jobs. There is a clustering of higher status jobs in the south and east of the country and a marked lack of them, particularly of jobs in the R&D of the new technologies, in the peripheral regions. For the peripheral regions this was seen to pose two problems: first, a lack of varied job opportunities for their inhabitants; and, second, and it was assumed consequently, the lack of social basis for future entrepreneurial and high tech growth. This interpretation of the nature of the problem was frequently accompanied by an, at least implicit, policy assumption that something was wrong with these regions, in their inherited economic and social structures perhaps, that could, through new forms of policy intervention, be righted in such a way as to enable them to emulate the proclaimed success of areas of the south and east at the other end of the country. In that light, much careful analysis, investigation, and prescription was undertaken: studies of the geography of innovation, of entrepreneurship, of the introduction of advanced technologies, and of new high tech products. Policy changes were formulated and, in some cases, pursued in an attempt to rectify the generally dismal results of these analyses: Numerous schemes evolved to promote local business people and local innovation; science parks were planned and, in some cases, established; suggestions were made for regional policy incentives to be differentiated by type of job.

This analysis, and the policy prescriptions that followed, are open to serious question. For the geography of jobs is only a spatial manifestation of the underlying geographical organization of production itself. Put at its crudest, the old sectoral division of labor between the geographical regions of the British economy has been, or is being, superseded by a division of labor between conception and execution. Most regions have jobs within the execution end of this spectrum, whereas the jobs in conception are clustered in only a few parts of the country. If this approach is followed through it has wide implications. It means that what the peripheral regions are lacking are not just high-status technical jobs, but the functions of control, coordination, and long-range strategic technological planning within the internal organization of industry. But, furthermore, there is only a limited range of such functions to be performed, particularly in a country no longer at the apex of the international division of labor. The new dimensions of interregional inequality are in the end no more than a particular spatial form of the changing national social and occupational structures outlined earlier in this chapter (which is not to say that geography is not important in the formation and molding of those national structures themselves, see Massey, 1984). That, in turn, has further implications both for the peripheral regions and for spatial policy more generally. It means, to begin with, that the "success" of areas such as the M4 corridor is quite simply not replicable in all those regions struggling to provide the conditions for its replication. Either there can be only one or two such regions or the jobs and functions within capitalist relations of production with which they are associated can be spread only thinly across the country. Yet, furthermore, there is considerable evidence, both economic and social, that the bundle of characteristics that go to make up these "top of the technology hierarchy" regions in themselves encourage clustering (see below and Massey, 1984). Yet if this is true, the economic success and the social status (at least for some) in these regions actually necessitates the economic drudgery, the failure of other regions to build such social and economic bases.

The spatial division of labor, in other words, depends on a technical division of labor within production. The collapse of the economies of the old engineering regions of the northwest and west midlands of England represents the decline of manufacturing industries in which the industrial technologist and scientist had a far more intimate and day-to-day relation to the shop floor, and in which the shop floor jobs were often themselves less routinized and more skilled than is true in the new technology industries that are now expanding. In these new industries, not only are conception and execution more separable than before in a functional sense, so too they are in a social and a geographical sense. There is, in other words, no way of seriously attacking such forms of spatial inequality without first attacking the social form of the organization of production, and the social divisions that appear to be endemic in much of the so-called new technology.

If the line of argument emerging (from some quarters) on serious consideration of interregional inequality thus concerns the dichotomized labor process, labor demand, and, consequently, social structure that appears to be associated with much of new technology, the argument emerging from the crippling unemployment levels of the inner cities concerns the products of new technology. For the inner cities there is little hope of attracting either the R&D end or the production end of employment in industries such as electronics. The exception here is London, especially in relation to new technology services, such as software production. In general, however, neither physical nor social environment is right for the high-status jobs (see below) and neither labor costs and organization nor physical infrastructure for production. And yet, within First World countries there is nowhere quite like the inner city for seeing, side by side, unused resources and unmet needs. Increasingly, the politically radical local governments that are at present in the ascendancy in many of Britain's major cities are pointing to the inadequacy of the profit motive, when left to work on its own, to bring the two things together. Increasingly, too, they are making the parallel argument about technology. As workers in London's Greater London Council and Greater London Enterprise Board point out, we may be able to play computer games or video television programs for consumption at a more convenient time, but the physically disabled within the capital city of one of the richest nations in the world still have to grope and stagger their way around, much as they have done for centuries. New technology could so easily provide the means to help, if only it were not so preoccupied with other things. Not only would the design and manufacture of such things provide socially useful products, they would also provide much needed jobs.

Urban and regional problems between them, then, are beginning to open up to question the whole shape of new technology, in terms of the internal division of labor and consequent social and spatial structures it seems to presuppose and in terms of the products it is used to produce. Finally, there is one other element, less tangible but equally significant in its potential social ramifications. This concerns some of the ideological constructions at present being built around new technology.

IDEOLOGIES AND NEW TECHNOLOGY

Scientists and technologists have a social image. In the United Kingdom the nature of that image has varied over the years, but for the whole of the post-World War II period it has contained the common idea that, one way or another, this relatively small group within the population is central to the economic health of the country. Scientists and technologists emerged from World War II with their status greatly enhanced (Burns and Stalker, 1961; Steward and Wield, 1984). Twenty years later Harold Wilson became prime minister on a platform in which he declared that Britain's bright new future would be forged in the White Heat of Technology. Today, the emphasis on individual entrepreneurship and inventiveness, the concern with relations between the academic and the business communities, the proliferation of such things as science parks help perpetuate, in suitably modified form, that image of centrality. The intermittent concern over "the brain drain" (the emigration of skilled workers, particularly scientists and technologists, and particularly up the hierarchy of the international division of labor) is evidence of this. Scientists and technologists are seen to be a key social group.

There are two ways in which this social power is important to the present analysis. The first concerns the construction and reformulation of ideologies of work. In particular, the labor process and social relations within the workplace that characterize the employment of these social groups within the labor market are held up as an example to all, as a prefiguration of what working life in the future could be like. A recent study of social relations and the organization of work in a high-status British science park revealed a number of important elements (Cochrane and Massey, 1984). Above all, image was important: the parkland setting, the status address, the clean working environment. (Indeed, en passant it is worth noting that in this preliminary investigation, the image of a university address—although only a "good" (i.e., high-status) university—was far more real in the eyes of those who chose to locate in science parks than was any working relationship between academic and business community.) Within this idyllic setting the organization of work could not be more different from, and is explicitly counterposed to, the conditions that most of the population daily faces in office, in mine, or on shop floor. In this classic, high-status science park, the social relations of work were characterized by individual flexibility between jobs; individual commitment to the company; frequently decentralized management structures; clean, bright, carpeted workplaces; individual interest in the job; voluntary flexibility in hours of work (image: staying on into the night, struggling over that knotty problem on the frontiers of science); and a total absence of trade unions.

This last, of course, is no accident. The whole construction is one that emphasizes individuality, individual commitment, and individual competitiveness, as opposed to collectivity and solidarity. It is an image that is used in the current attack on trade unionism, as trade unionists fight a losing battle to save the crumbling sections of the economy that have long been their bases. Flexibility of jobs is counterposed to the unions' job demarcations and "restrictive practices"; commitment to the company is counterposed to affiliation to trade unions organized on an intercompany basis; the clean workplaces to the antiquated factories that are the legacy in so many parts of British industry and that the labor movement is now forced, through need for jobs, to defend; staying late voluntarily is counterposed to the endless disputes over overtime. It appears to be a powerful case that can easily put trade unionists on the defensive. Indeed, it does put them on the defensive, but is it really so powerful?

The critique of this position contains many of the same elements as emerged from the critique of regional policy. They are based around the nature of the division of labor that at the moment seems to be associated with new technology, and around the nature of the social mechanisms that allocate different groups within society to different places within that division of labor. Together, these mean that this is a form of work organization that, by its very nature, cannot be extended to all; indeed, it can have negative effects for those in other parts of the social division of labor.

The employees in the science park, and in most of the higher status jobs within new technology as it is at present defined, are middle class, college graduates, white, and male. They are high-status employees, a fact of which they are well aware. But status can only be high if it can be counterposed to low. These characteristics of work organization show no signs at all of spreading to those working in, for instance, the mass-production end of new technology industries. Indeed, to some extent, by identifying trade unions as old fashioned, they are being used to undermine what sources of protection there have been for workers in these parts of new industries. The new attitude to work also goes against many demands currently finding their way onto the agenda of the labor movement. Staying late at work on that fascinating problem is all very well for the male worker; but who, meanwhile, is holding the fort back home? On whose extra labor does this commitment to the company depend? And whose life and timetable has to be reorganized as a result? This ideology puts paid work above the rest of life, makes it central to life and, indeed, to the employee's self-image in a way that many, in particular those in the women's movement, have long criticized. It goes, too, against the current struggles in the context of mass unemployment, while those with jobs see their work intensified for a shorter working week.

Second, it was mentioned above that these high-status high tech jobs are concentrated in a few parts of the United Kingdom. There are economic reasons for this: linkages of various sorts, the highly individualized nature of the labor market, particular advantages such as proximity to airports, motorways, and defense establishments. But there are also social reasons. There is now a cachet attached to the often semirural parts of the country in which these sections of the labor force live. And these people are probably the only ones in the United Kingdom who can choose to live where they wish in the knowledge that jobs will follow. Together, all these factors produce a geographical clustering (see Massey, 1984; for a similar analysis of the United States, see Dorfman, 1983). It is a geography that, both by the sense of status attached to living in the "right" area and by the relative separation of these parts of the country from old-fashioned production, reinforces the social distance between different parts of the labor force.

There have, in the past, been critiques of the organization and production of science and technology. There was fierce debate within the scientific community itself (Steward and Wield, 1984) in the period around and after World War II. But these critiques, at least those that gained prominence, were primarily concerned with the nature of control over technology and, to some extent, the nature of its products. The present critique goes deeper, to question the direction of technological development itself. The Wilson period, in part precisely because it was an era of Labour Government, has come under particular attack. The "white heat of technology" in practice came to mean an emphasis on size, top-down planning, the expansion of the professional and scientific strata that controlled the new technology, the expansion of concomitantly deskilled jobs, and, combined with a regional policy that operated in terms only of numbers of jobs rather than quality, the spatial/regional dichotomy referred to above. It was in the 1960s and the first half of the 1970s that this geographical pattern, a new spatial division of labor, was most strongly established. But, above all, the policies of the Wilson era assumed that the direction of technological development was neutral. "New technology" was "out there," waiting to be discovered.

ALTERNATIVES

In one of many telling phrases, Mike Cooley, a technologist who once worked at Lucas Aerospace and who now works at the Greater London Enterprise Board (GLEB), the Greater London Council's Economic Strategy arm, observed that new technology is not "out there," like a new continent waiting to be discovered. New technology is created; and it is a social creation. Our responsibilities do not lie solely in mitigating its effects, in adapting society to its demands and implications. There is a social choice also, and a social responsibility, for the very nature of new technology.

The experiment at Lucas Aerospace has now become world famous (Wainright and Elliot, 1982). The plan arose in the face of threatened lay-offs and was designed by a shop stewards combine. It proposed alternative products, alternatives to the firm's normal output of military-related equipment, and alternative ways of using the skills of the employees whose jobs were threatened. An impressive range of socially useful products was developed. But there was much more to it than new products. There was also an attempt to rethink how those products should be made. The nature of the labor process and the skill-hierarchized division of labor were reorganized, and products were developed in consultation with potential consumers (patients and workers in the health service, for example). The Shop Stewards' Plan was not adopted by the company; but it became a symbol of what might be possible. A range of other experiments came to prominence. One such experiment, important in the discussion here, was the development of "human-centered" technology; that is, high technology tools that leave control and skill with the operator.

It is on these kinds of ideas that some local authorities in the United Kingdom are now trying to build. Faced with devastating economic problems, there is a growing awareness that new technology, in its currently accepted definition, in itself offers little way out. As developed and used at the moment, it does not provide the jobs and services that are lacking; if it provides any jobs it is likely to provide few, certainly only a few "quality" jobs, and there is a growing awareness that the quality of work, in the widest sense (both what is made and how it is made) is as big a question for the future as is the number of jobs.

I will concentrate here on one example of these experiments: that of the GLC and GLEB. It is probably at the present time the best-known and the most ambitious. Under the general rubric of "restructuring for labor" (as opposed to capital), three working principles were derived to govern the GLC's overall industrial strategy. These were as follows: (1) the principle of bringing wasted assets—human potential, land, finance, technological expertise, and resources—into production for socially useful ends; (2) the principle of extending social control of investment through social and cooperative ownership and increased trade union powers; and (3) the principle of development of new techniques that increase productivity while keeping human judgment and skills in control.

All three principles in one way or another relate to technology; technological innovation is fundamental to principles 1 and 3.

But the GLC's Industry and Employment Committee does not endorse the view that new technology is somehow inherently progressive, or that there is even such a thing as *the* new technology. Rather they recognize that there are several different directions in which technology could develop, according to different social and economic objectives. The Industry and Employment Committee has defined its objectives in a general way but the implications for the choices of technology will be developed through the popular planning process. . . . For this to be possible facilities and expertise on technological matters need to be made far more accessible to working class people who in the past have been excluded from judgement between different technologies (Capital and Class, 1982: 125-126).

The criteria for socially useful production and technology have been defined as follows:

Socially useful products are such as to conserve energy and materials (both in manufacture and in use); their manufacture, repair and recycling are carried out by nonalienating labour; and the production and the products themselves should assist human beings rather than maiming them (Capital and Class, 1982: 126).

One of the ways in which an attempt is being made to feed these principles about new technology into the local authority employment strategy is through the development of "technology networks." The aim of the networks is that they should

harness the often underused facilities of London's polytechnics and universities, for the use of trade unionists and others who wish to develop employment plans either as bargaining positions in their own company or as the basis of a co-operative or municipal enterprise. The networks will for instance develop a "product bank" of prototypes of new products of a socially useful kind that would be available for such groups (Capital and Class, 1982: 126).

Some of the networks are based on product areas, such as energy or transport; others have an area basis (e.g., northeast London). One technet (the London New Technology Network, LNTN) is concerned specifically with new technology. A range of examples of its work was recently detailed in the compaigning newspaper "Jobs for a Change" (Number 11, June, 1984). One of its key exemplary projects is the development of "expert systems" that enable medical diagnoses normally performed by consultants to be done by the local general practitioner. It is at present being used for treatment of diabetes and clearly has a wide range of applications. It is, in other words, an attempt to enable democratization of medical knowledge. Another project is aimed at developing systems of small, inexpensive computers to be linked up in such a way that information can be shared among different locally based organizations within a community. LNTN is also importantly engaged in the further development of human-centered technologies, ranging through a variety of tools, computer graphics, and robotics.

It would be wrong to give the impression that these experiments are not facing problems. The problems are in fact enormous, and they are at least as much human and social as technological. Probably the main problem is convincing people to have the confidence and interest to join in. Most people are so accustomed to not having control over technology that they assume they have no technological skills or abilities. In this sense, to achieve anything like their full objectives technology networks are necessarily a long-term project (they have only been going for between one and two years). There is frequently overoptimism about the ease with which human-centered alternative technologies can be developed. There is sometimes a danger of even these products being taken over by commercial uses in such a way as to turn them against their original aims. Finally, these are very small-scale and piecemeal experiments. GLEB's total annual funding is only £30m and its future is currently in doubt due to the proposed abolition of the GLC (the government of London) by the national government. But these experiments are not the only ones under way. There are others within the United Kingdom and also elsewhere in Europe (Dickson, 1984), the science shops in Holland perhaps being the best known. It is easy to decry the smallness of these initiatives. But what if we were to pour into them the same kinds of resources as are currently poured by every Western industrial government, and by thousands of companies, into the presently dominant form of new technology?

Even in their present experimental and limited form, what these initiatives demonstrate is that new technology is not out there to be discovered. Alternative futures exist depending on the way in which technology is developed, directed, controlled, and used. On all these aspects there are choices. Instead of simply predicting new technology and coping with its social implications it could be designed and controlled from the beginning to have beneficial social implications. Were some of the present experiments ever to become the dominant form of new technology, we would not be facing today the problems of massively deskilled jobs, of dichotomized divisions of labor, and similarly dichotomized social structures. Nor would we be facing the particular kinds of regional problems that we face today in the United Kingdom. And as all these problems, in some mediated form or another, are based on the organization of the social relations of production, it is indeed difficult to see how they can be addressed without addressing also the internal organization of production—and that includes the form that new technology should take.

REFERENCES

BURNS, T. and G. M. STALKER (1961) The Management of Innovation. London: Tavistock.

Capital and Class (1982) "Strategy: a socialist GLC in capitalist Britain?" 18 (Winter): 117-133.

COCHRANE, S. and D. MASSEY (1984) " A new ideology of work?" Presented at the Annual Conference of the Conference of Socialist Economists, CSE, London.

DICKSON, D. (1984) " 'Science shops' flourish in Europe." Science 223 (March 16): 1158-1160.

DORFMAN, N.S. (1983) "Route 128: the development of a regional high technology economy" Research Policy 12: 299-316.

MASSEY, D. (1984) Spatial Divisions of Labour: Social Structures and the Geography of Production. London: Macmillan.

STEWARD, F. and D. WIELD (1984) "Science, planning, and the state." Open University Faculty of Technology. (mimeo)

WAINWRIGHT, H. and D. ELLIOTT (1982) The Lucas Plan: A New Trades-Unionism in the Making? London: Alison and Busby.

About the Contributors

BARBARA BARAN is a researcher at the Institute of Urban and Regional Development at the University of California, Berkeley and a doctoral candidate in City and Regional Planning there. She currently is researching office automation and the restructuring of services under the impact of high technology.

ROBIN BLOCH currently is working on a doctorate degree in Urban and Regional Planning at University of California, Los Angeles. From Johannesburg, South Africa, he has studied at the Universities of Cape Town, Witwatersrand, and California at Berkeley. He is studying urban and regional development in South Africa and the United States.

MANUEL CASTELLS is Professor of City and Regional Planning at the University of California, Berkeley. He received his Ph.D. in Sociology from the University of Paris in 1967, with a dissertation on the technological factors of industrial location in the Paris region. He taught at the University of Paris for 12 years, as well as at the Universities of Montreal, Chile, Wisconsin, Copenhagen, Boston, Mexico, Hong Kong, Southern California, and Madrid. He has published 12 books, including *The City and the Grassroots,* winner of the 1983 C. Wright Mills Award.

CLAUDE S. FISCHER is Professor of Sociology at the University of California, Berkeley. His most recent publications are *To Dwell Among Friends: Personal Networks in Town and City* (Chicago, 1982) and *The Urban Experience* (2nd ed., Harcourt Brace Jovanovich, 1984). He currently is conducting research on the diffusion and consequences of new transportation and communications technologies in the first half of the century.

AMY K. GLASMEIER is Assistant Professor of Economic Planning at the Pennsylvania State University. She obtained her Ph.D. in City and Regional Planning from the University of California, Berkeley, with a dissertation on high technology industries and regional development.

PETER HALL is Professor of City and Regional Planning at the University of California, Berkeley and Professor of Geography and Chairman of the School of Planning Studies at the University of Reading in England. He is the author of *London 2000* (Faber & Faber, 1963), *The World Cities* (St. Martins, 1984), *Urban and Regional Planning* (Penguin, 1982), and *Great Planning Disasters* (University of California Press, 1982). His research interests center on the geography of high technology industry.

LARRY HIRSCHHORN is a Senior Researcher at the Management and Behavioral Science Center, Wharton School, University of Pennsylvania. He is the author of *Beyond Mechanization: Work and Technology in a Post-Industrial Age* (MIT Press, 1984), a study of modern automation and work design. He consults for government and industry on such issues as organization process, job design, and planning. He currently is working on the relationships between information technology and professional work. He also is doing a study of the psychodynamics of work.

ANN ROELL MARKUSEN is an economist and Associate Professor of City and Regional Planning at the University of California, Berkeley. She has written widely on regional economics, industry studies, and politics. Her books—*Profit Cycles, Oligopoly, and Regional Development* (MIT Press), *The Politics of Regions* (Rowan & Allenheld), and *Silicon Landscapes* (with Peter Hall, Allen & Unwin)—all deal with aspects of high technology and regional development. She presently is engaged in research on the relationship between military spending and high technology development.

DOREEN MASSEY is Professor of Geography in the Faculty of Social Sciences at the Open University, U.K. She is the author of *Spatial Divisions of Labour* (Macmillan) and (with Richard Meegan) of *The Anatomy of Job Loss*. She is a Director of the Greater London Enterprise Board.

LIONEL NICOL is a lecturer in the Infrastructure Planning and Management Program of the Department of Civil Engineering at Stanford University. He also is a Research Associate in the Communications and Development Program at the Institute of Urban and Regional Development, University of California, Berkeley. He holds a Masters degree in City Planning and an M.A. in Economics from the University of California, Berkeley, where he is working on his doctoral dissertation. His research focuses on the effects of new information technology on national and regional economic development. He is involved in several projects with international organizations to evaluate the economic impacts of telecommunications investments in developing countries.

FRANÇOISE SABBAH is a professional journalist who holds a Masters Degree in Journalism from the University of Strasbourg, a Doctorate in Sociology from the University of Paris, and a Masters in Communication Arts from San Francisco State University. She currently is a Technical Advisor to the President of the Government, Madrid, Spain.

ANNALEE SAXENIAN is a Research Associate in the Program in Science, Technology, and Society at MIT, where she is completing a doctorate degree in Political Science. Her dissertation examines the politics of the microelectronics revolution in the United States, in particular, the organization and politicization of the Silicon Valley business community and the emergent ideology of high technology. She has written extensively on the urban and regional consequences of high tech industries, a research agenda that grew out of her Master's thesis in the Department of City and Regional Planning at the University of California, Berkeley.

THOMAS M. STANBACK, Jr., is a Senior Researcher at the Conservation of Human Resources Project, Columbia University and a Professor of Economics at New York University. He has published several major books, including *Understanding the Service Economy* (1979), *Services: The New Economy* (1981), and *The Economic Transformation of American Cities* (1984, with Thierry Noyelle). His research interests include urban workforce changes and technology and changing employment patterns.

MICHAEL STORPER is Assistant Professor of Urban Planning at University of California, Los Angeles. He studied Geography at

Berkeley prior to his arrival in Los Angeles. He is working in several areas, including the role of spatial organization in shaping technological innovation and macroeconomic growth in the United States during the last 100 years; the changing organization and location of the motion picture industry, and its implications for understanding metropolitan labor markets; and the macroeconomic implications of decentralization policies applied to Third World metropolises.

RICHARD A. WALKER is Associate Professor of Geography at the University of California, Berkeley. He received his doctorate degree from Johns Hopkins University, under the direction of David Harvey, in 1977. He has written on a diverse range of topics in geography, including urban history, environmental policy, philosophy, and industrial location theory. He also uses his wit in public affairs, currently as coordinator of the Faculty for Human Rights in El Salvador and Central America.